Biotechnology A Guide to Genetic Engineering

Contents

PREFACE		vii	
INTRODUCTION		What is Biotechnology	xi
CHAPTER	1	DNA: The Molecule of Life	1
CHAPTER	2	DNA: The Messenger of Life	23
CHAPTER	3	When Something Goes Wrong	41
CHAPTER	4	Isolating and Manipulating DNA	54
CHAPTER	5	Preparing a Portrait: The Characterization of a DNA Molecule	72
CHAPTER	6	Cloning a Gene: How to Photocopy Heredity	92
CHAPTER	7	Cloning–Part Two: Using a Gene Library	116
CHAPTER	8	Factors That Govern Gene Expression	139
CHAPTER	9	Cloning Genes in Yeast Cells	160
CHAPTER	10	Cloning Genes in Mammalian Hosts	170
CHAPTER	11	Applications of Biotechnology: Human Therapeutics	192

CHAPTER 12 Applications of Biotechnology: Agriculture 220

FURTHER READINGS 233

GLOSSARY 235

CREDITS 242

INDEX 243

Preface

Few aspects of our daily lives remain untouched by the developments of biotechnology. In the past decade-and-a-half, scientists gained the ability to peer into the innermost workings of hereditary material. Using this information researchers have analyzed the structures and functions of a great many human and nonhuman genes and now understand many of the details of the transfer of hereditary information from DNA and RNA to polypeptide product. Equally important, scientists can now manipulate the transfer of hereditary information through this pathway. The intimately paired sciences of biotechnology and genetic engineering have grown through the accumulation of such knowledge, resulting in enormous changes in our everyday existence. Areas as diverse as diagnostic medicine, preventative medicine, animal husbandry, agriculture, and predictive medicine have been immeasurably affected.

Despite the undeniable impact of biotechnology and genetic engineering on our lives, one commonly encounters people—both students and poststudents—who know little or nothing about these profoundly important sciences. A common perception exists that science, in general, and biotechnology and genetic engineering, in particular, are difficult if not impossible to understand. Yet I contend that these difficulties stem from the manner in which the subjects traditionally are presented to students. All too often, science texts are enormous tomes filled to the brim with facts and figures presented in a dry and unreadable fashion. It is not surprising if a student, after an experience with such a text, comes away feeling that the science itself is a dry and unpleasant, if not impossible, subject. Yet we find a paradox: the various aspects of biology, including biotechnology and genetic engineering, describe life itself. What should be more fascinating to the student?

This book attempts to present a readable and understandable text that takes the student from a basic understanding of genetic structure and function all the way to an understanding of the basics of biotechnology and genetic engineering. As such, it is appropriate for college and university students with a variety of science backgrounds as well as for the highly motivated high school student and the student enrolled in a seminar or institute-type training course. This text, a self-contained unit, begins with information describing DNA and RNA and the flow of hereditary information, and proceeds through analyses of the results of changes in hereditary information. A series of chapters describes the manipulation of genetic material to meet a research requirement, and the final two chapters discuss the impact of these genetic manipulations on various aspects of our lives.

Perhaps the most difficult part of writing this text has been to decide what information to include. To fit in with the goal of presenting a "user-friendly" and easily understandable text, I ruthlessly pruned the vast amount of available material and sometimes sacrificed descriptions of specific techniques in favor of covering the theory and background of biotechnology and genetic manipulation. I hope that the interested student will consult one of the many advanced texts available in the subject and fully understand the information presented there.

AIDS TO THE STUDENT

Chapter previews and overviews consisting of relevant anecdotal material and questions provide a brief insight into the information covered in each chapter. The chapters contain the information necessary to allow the student to answer the questions posed in the chapter overview.

Key terms, shown in bold letters to visually distinguish them from the body of the text, are those terms presented and defined for the first time, or which represent essential concepts which are new to the reader.

The *Review and Summary* section for each chapter condenses and restates the main ideas covered within the chapter. Each chapter's Review and Summary, presented in list form, provides a convenient and effective review of covered material. In addition, the chapter Review and Summary is useful for students wanting to preview material prior to beginning study of each chapter and later provides useful reinforcements of materials to be reviewed for testing.

Offset materials provide in-depth descriptions of pivotal techniques and instruments which are beyond the scope of the main text itself. While not essential to an understanding of biotechnology and genetic engineering, these techniques and instruments are common to many working laboratories and are therefore of practical interest.

Additional readings are suggested for students interested in more detailed information on specific topics. Readings are chosen primarily from science publications written for the educated layperson rather than from technical science journals.

SUPPLEMENTARY MATERIAL

An *Instructor's Manual* offers detailed outlines and chapter summaries as well as additional supplementary material and a bank of test questions for each chapter. The supplementary materials consist of additional background information to complement material presented in the student text, in depth discussions of techniques, and other pertinent information not included in the main text.

ACKNOWLEDGMENTS I would like to thank my husband Mark, and children, Becky, Jonah, and Hannah for their remarkable patience with this project. I would also like to thank my editor Meg Johnson, for her assistance in the preparation of this text. In addition, I would like to thank the following individuals for support and encouragement during the final push as this text was being completed in the spring of 1991: Rebecca Donian, Maria Finnegan Hall, Sandra MacKenzie, Luanne Meyer, Kathe Scherba-Nielsen, and Laura Weston. I would also like to thank Elaine Chesebrough for her help with word processing and Debbie Glaister for many helpful technical discussions.

I would also like to thank Marilyn J. Tufte, University of Wisconsin, Platteville; David J. Betsch, Director, Biotechnology Training Programs; Margaret A. Tubbert, Onondaga Community College; G. Rickey Welch, University of New Orleans; Edward C. Kisailus, Canissius College, and Sanford I. Bernstein, San Diego State University.

Introduction: What is Biotechnology?

Biotechnologists of one sort or another have flourished since prehistoric times. When the first human beings realized that they could plant their own crops and herd their own animals, they learned to use biotechnology. The discovery that fruit juices fermented into wine, or that milk could be converted into cheese or yogurt, or that beer could be made by fermenting solutions of malt and hops began the study of biotechnology. When the first bakers found that they could make a soft, spongy bread rather than a firm, thin cracker, they were acting as fledgling biotechnologists. The first animal breeders, realizing that different physical traits could be either magnified or lost by mating the appropriate pairs of animals, engaged in the manipulations of biotechnology.

What, then, is biotechnology? The term brings to mind many different things. Some think of developing new types of animals. Others dream of an almost unlimited source of insulin to treat people with diabetes. Still others envision the possibility of growing crops that are more nutritious and naturally pest-resistant to feed a rapidly growing world population. This question elicits almost as many first-thought responses as there are people to whom the question can be posed.

In its purest form, the term **"biotechnology"** refers to the use of living organisms or their products to enhance human health and the human environment. Prehistoric biotechnologists did this as they used yeast cells to raise bread dough and to ferment alcoholic beverages, and bacterial cells to make cheeses and yogurts and as they bred their strong, productive animals to make even stronger and more productive offspring.

Throughout human history, we have learned a great deal more about the different organisms that our ancestors used so effectively. The marked increase in our understanding of these organisms and their cell products gains us the ability to control the many functions of various cells and organisms. Using the techniques of **gene splicing** and **recombinant DNA technology** we can now actually combine the genetic elements of two or more living cells. Functioning lengths of DNA can be taken from one organism and placed into the cells of another organism. As a result, for example, we can cause bacterial cells to produce human molecules. Cows can produce more milk for the same amount of feed. And we can synthesize therapeutic molecules that have never before existed.

Figure I.1
Gregor Johann Mendel studied the inheritance of physical traits in pea plants. Using information gathered in these studies, Mendel elucidated the basic principles of heredity.

The seeds of the modern biotechnological revolution were first planted in 1865 when Gregor Mendel (Figure I.1) published papers that described his research into the inheritance of seven different physical traits in the common garden pea. This body of research eventually led to the concept of the gene as the basic unit of heredity. Over the next century, other research necessary for the growth of modern biotechnology slowly accumulated (Table I.1).

As is shown in Table I.1, the phenomenal growth of biotechnology as an industry began in the late 1970s. A group of researchers realized that the delicate genetic manipulations, which were becoming increasingly possible, might have commercial applications. Some of these scientists isolated the gene which encodes human somatostatin, a human hormone consisting of only 14 amino acids, and used the appropriate enzymes to insert it into a cloning vehicle. That cloning vehicle was, in turn, inserted into a bacterial cell which, as a result, began to synthesize the human somatostatin hormone. Scientists had created a bacterial cell with the ability to synthesize a human hormone!

The few years since this remarkable development have seen incredible growth in biotechnology. Insulin, growth hormones, and interferons have been cloned. Vaccines have been synthesized. Some of the molecules that control and modulate the complex immune response have been isolated and cloned. Large quantities of novel therapeutic molecules necessary for both therapeutic use and research have been synthesized and many can now be produced. Strains of bacteria have been developed which have the ability to ingest and break down oil. Self-fertilizing, insect-resistant plants have been produced, as well as cereals and grains with higher-than-average nutritive values. The potential to identify each and every human gene now exists as the intricate chromosomes of the human genome have begun to yield to attempts to map them. These are only a few examples of the scientific insights which have developed as a result of biotechnological research.

Table I.1

1866	Gregor Mendel presents research leading to the concept of the gene as the basic unit of heredity.
1869	Friedrich Miescher (Figure I.2) isolates nuclein, later shown to be DNA, from the nucleus of a white blood cell.
1885	*E. coli* bacterial cells are identified and grown under controlled conditions.
1902	Archibald Garrod's observation that the human disease known as Alkaptonuria behaves as a Mendelian recessive trait leads to the suggestion that enzymes are encoded by genes.
1910	Thomas Hunt Morgan (Figure I.3) provides the first evidence that genes are contained on chromosomes.
1940	George Beadle and Edward Tatum hypothesize that one gene encodes one enzyme.
1944	Oswald Avery and colleagues demonstrate that DNA, not protein, carries hereditary information.
1952	Alfred Hershey and Martha Chase demonstrate that the genes of a bacteriophage are made of DNA and are capable of directing the synthesis of new bacteriophage proteins.
1953	James Watson and Francis Crick (Figure I.4a) draw on the work of Rosalind Franklin (Figure I.4b) and others to propose that the DNA molecule is an alpha double helix structure in which the two strands are both complementary and antiparallel to one another.
1955–65	The role of tRNA, mRNA, and rRNA, as well as of DNA and RNA polymerases in gene function is elucidated.
1957	The Central Dogma, which states that hereditary information flows from DNA to RNA to protein, is put forth by Francis Crick and George Gamov.
1961	Marshall Nirenberg and Har Gobind Khorana correctly translate the genetic code.
1967	DNA ligase is isolated and identified.
1970	Stewart Lin and Werner Arber identify the first restriction endonucleases.
1970	Researchers discover the enzyme reverse transcriptase, which catalyzes the reaction in which DNA is transcribed from an RNA template.
1972	Paul Berg creates the first recombinant DNA molecule.
1973	Stanley Cohen and Herbert Boyer and colleagues construct a functioning plasmid which contains genes which confer resistance to both tetracycline and streptomycin.
1977	The first commercial biotechnology companies are formed.
1977	Somatostatin becomes the first human hormone to be synthesized by a bacterial cell as a result of transformation with human DNA.
1977–present	The field of biotechnological research sees exponential growth.

Figure I.2
Friedrich Miescher discovered "nuclein," later shown to consist primarily of DNA, in cellular nuclei.

Figure I.3
Thomas Hunt Morgan, working with the fruitfly *Drosophila melanogaster*, provided the first evidence showing that genes are contained on chromosomes.

This textbook seeks to explain the tools of biotechnology. Chapters one through three focus on the basic material of biotechnological research. That material is, of course, deoxyribonucleic acid—more commonly known as DNA. Chapter one examines the structure of this very important molecule, while chapter two explains the functions of the DNA molecule and how those functions are irrevocably intertwined with another type of nucleic acid known as ribonucleic acid, or RNA. Chapter three explores the results of "mistakes" in the genetic code of the DNA molecule.

A B

Figure I.4
(a) James Watson (left) and Francis Crick, using data provided from a variety of laboratories including the X-ray diffraction data of (b) Rosalind Franklin and others, elucidated the structure of the DNA molecule.

Chapters four and five, respectively, cover the techniques of DNA manipulation and DNA characterization. Without the ability to manipulate and characterize DNA molecules with great accuracy, modern biotechnology would be nothing but an impossible dream. Chapter six begins the discussion of gene cloning—the basic technique of biotechnology—with the steps involved in the cloning of any gene and the production of a gene library. In addition, it discusses the use of prokaryotic cells as hosts in the gene cloning process. Gene libraries and their uses are covered in some detail in chapter seven. Chapter eight concerns the various parameters which control the expression of cloned genes, and the organization of both prokaryotic and eukaryotic genes. Chapters nine and ten discuss the use of yeast and higher eukaryotic cells, respectively, as hosts in the cloning process. Finally, chapters eleven and twelve present some of the applications of modern biotechnology.

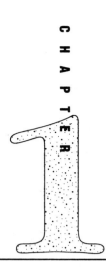

DNA: The molecule of life

In 1981 a few young men died of an old man's disease. Their deaths set the stage for one of the most important and intriguing medical detective stories of modern medicine.

These men died from a rare type of cancer known as Kaposi's sarcoma. Even more surprising, several of these young men had also suffered from an unusual type of pneumonia, a type so rare that only a handful of cases had been diagnosed in the previous twelve years.

As doctors pondered the mystery of why otherwise healthy young men should be stricken with not one, but two, very rare diseases, they also began to notice an increase in the number of patients with swollen lymph nodes and with another type of cancer called undifferentiated non-Hodgkins lymphoma.

In 1982 this syndrome—pneumonia, Kaposi's sarcoma, swollen lymph nodes, and lymphoma—was given the name Acquired Immune Deficiency Syndrome, or, more commonly, AIDS. By the end of 1982, nearly one thousand men and women had died from AIDS. Yet scientists still didn't know what was causing the deadly disease.

The breakthrough came in late 1982. Scientists working with Drs. Luc Montagnier in Paris and Robert Gallo in Washington, DC, realized that a retrovirus—a special type of virus containing RNA as its genetic material—most likely caused the deadly disease. Once a retrovirus invades a cell, enzymes use the viral RNA as a template to synthesize a complementary molecule of DNA. Then the invaded cell synthesizes many more virus particles using the directions specified by the genetic information stored in the newly synthesized DNA molecule. In most cases, the new virus particles eventually kill the invaded cell.

To understand how the AIDS virus, or any virus, wreaks such havoc, one must understand how the relatively simple structure of the DNA molecule stores genetic information.

DNA: The Molecule of Life

Figure 1.1
The diversity of life forms. It is perhaps surprising that relatively similar DNA molecules can lead to life forms as diverse as (a) human children, (b) the slime mold *Dictyostelium discoideum,* (c) Canadian geese, and (d) the giant redwood tree.

OVERVIEW About five billion people live on Earth today, and except for identical twins, no two of them are alike. Yet every one of those living human beings has DNA of the same basic structure. Even more amazing is that the DNA in the cells of *your* body closely resembles the DNA in the cells of any other organism—a Canada goose, a giant redwood tree or even slime mold (Figure 1.1)! This striking variability of life forms results from the remarkable variety of information that can be contained in the simple structure of the DNA molecule.

DNA: The Molecule of Life

C

D

The flow of hereditary information from one generation to another relies on the simple, yet elegant, structure of the DNA molecule. In this chapter we will examine this structure and see how DNA is able to contain all of the information needed to account for genetic diversity. Consider the following questions:

1. What cellular component holds the blueprints needed to construct the whole organism?
2. What are the building blocks of a DNA molecule and how are they put together?
3. What special features of DNA structure allow the molecule to replicate so accurately?

THE STORAGE AND TRANSMISSION OF HEREDITARY INFORMATION

"Factors" that Direct Physical Characteristics

We all know that members of the same family tend to resemble one another. A new baby may have dad's nose, mom's eyes, and Uncle Charlie's red hair. We take these resemblances for granted. Yet how can we explain them?

In the mid 1800s a monk named Gregor Mendel, working in Brunn, Austria (now known as Brno, Czechoslovakia), carried out an amazing piece of scientific detective work. Mendel observed that the offspring of certain plants had physical characteristics similar to the physical characteristics of the plants' parents or ancestors. Mendel wondered why related organisms, both plant and animal, tended to resemble one another and how familial resemblances might be explained. Mendel reasoned that close observation of inheritance might provide him with the answer for which he searched. He therefore set out to examine and quantify the physical traits in pea plants in an attempt to predict the traits that would occur in future generations. During years of painstaking work, Mendel counted many thousands of instances of seven different traits, including plant height, flower color and seed shape (Figure 1.2). Mendel concluded that certain particles or "factors" were being transmitted from parent to offspring and so on, thus providing a connection from one generation to the next. Mendel suggested that these factors were directly responsible for physical traits. His interpretation of the experimental data further suggested that each individual had not one, but two factors for each trait, and that these factors interacted to produce the final physical characteristics of the individual. Both the location and the identity of Mendel's factors remained unknown for many years.

THE NUCLEUS—STOREHOUSE FOR HEREDITARY INFORMATION

In 1943 a Danish biologist named Joachim Hammerling carried out an important experiment in which he searched for the part of a cell that directs its physical appearance or **phenotype.** Hammerling used large unicellular green algae called *acetabularia* (Figure 1.3). Each individual of the *Acetabularia* species is composed of one single, large cell about 6 cm long. Each cell has three main body

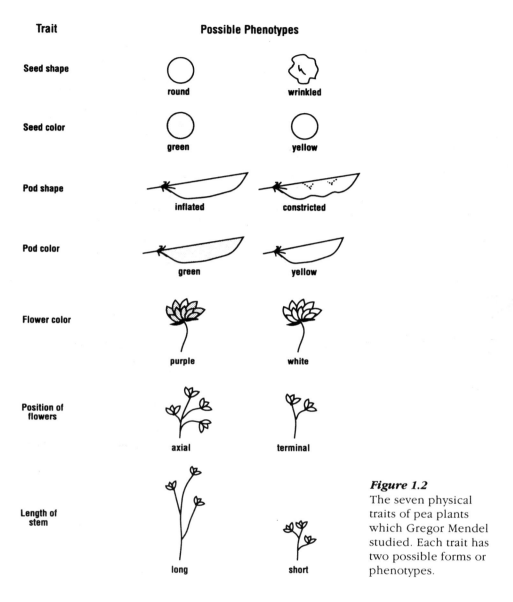

Figure 1.2
The seven physical traits of pea plants which Gregor Mendel studied. Each trait has two possible forms or phenotypes.

parts: a foot or **base** contains the nucleus and anchors the cell to a rock or other support, a **stalk** resembling a plant stem, and a **cap** which carries out the process of photosynthesis (Figure 1.3a). Hammerling used two different species of *Acetabularia*. One of the species, *Acetabularia mediterranea,* has a disk-shaped cap while the other species, *Acetabularia crenulata,* has a branched cap, more like a flower. Hammerling cut the stalk and cap off of an *A. mediterranea* cell and grafted a stalk from *A. crenulata* in its place. He then asked the question: What type of cap—disk or branched—will grow on the *A. mediterranea* base? The answer was that a modified branched cap, with marked similarity to *A. crenulata* caps, grew. Hammerling then removed the new cap from the *A. mediterranea* base to see what type of cap would now grow on the grafted stalk. This time a disk-shaped cap, just like the *A. mediterranea* caps, grew (Figure 1.3b).

A

Figure 1.3
Hammerling's *Acetabularia* experiment showed that the nucleus of a cell contains the genetic information which directs cellular development. (a) *Acetabularia* is a single-celled green alga. *A. mediterranea* has a smooth disk-shaped cap while *A. crenulata* has a branched flower-like cap. Each *Acetabularia* cell is divided into three segments consisting of the "foot" or "base" which contains the nucleus, the "stalk," and the "cap." (b) In his experiments, Hammerling grafted the stalk of one species of *Acetabularia* onto the foot of another species. In all cases, the cap that eventually developed on the grafted cell matched the species of the foot rather than that of the stalk.

This experiment suggested to Hammerling that the factor directing the growth of the algal cap was located in the base of the cell. He felt that the growth of the modified branched cap resulted from the presence of a message of some sort in the grafted stalk. That message was used up in directing the production of the modified *mediterranea* cap so that when a second cap was regenerated, it grew according to instructions derived from the nucleus contained in the base of the cell.

Hammerling concluded from these experiments that the nucleus of the cell was both directing its development and somehow specifying its hereditary characteristics. Later experiments by other scientists confirmed this conclusion and broadened its implications to suggest that not only did the nucleus direct the synthesis of new parts of an individual cell, but also that the nucleus directed the growth and development of entire multicellular organisms. These experiments reinforced and expanded data gathered by the microscopic analysis of cellular nuclei which had begun in the 20th century, and which is described below.

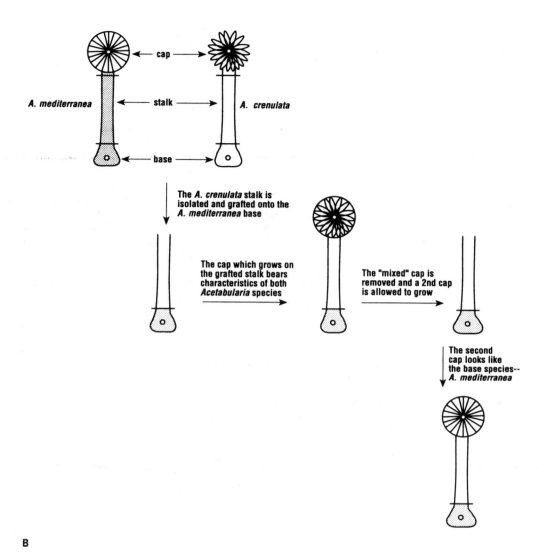

B

NUCLEAR HEREDITARY MATERIAL IS FOUND ON CHROMOSOMES

The identity of Mendel's factors had remained unrecognized until the turn of the century, some forty years after Mendel's painstaking experiments. At that time two exciting scientific developments came together, allowing scientists to actually see the material found inside of the cell's nucleus. These two developments were the construction of increasingly powerful microscopes and the discovery of dyes or **stains** which selectively colored the various components of the cell. As scientists examined cellular nuclei they observed long, thin, rod-like structures which tended to become colored when the cell was treated with certain stains. They called these nuclear structures **chromosomes.** Many more microscopic observations confirmed these facts:

1. A variety of chromosome types, as defined by relative size and shape, are present in the nucleus of each cell. Furthermore, there usually were two copies of each type of chromosome. This situation is referred to as **diploidy** and the cell is called a **diploid cell.**

2. All of the cells of an organism, excluding sperm cells, egg cells, and red blood cells, and all organisms of the same species have the same number of chromosomes.

3. The number of chromosomes in any cell doubles immediately prior to the cell division processes of **mitosis** and **cytokinesis,** in which a single cell splits to form two identical offspring cells.

4. The sex or **germ** cells (e.g., sperm and egg) have exactly half of the number of chromosomes as are found in the non-germ or **somatic** cells of any organism. Furthermore, the germ cells have just one copy of each chromosome type. Such cells are called **haploid cells.**

5. The fertilization of an egg with a sperm cell produces a diploid cell called a **zygote,** which has the same number of chromosomes as the somatic cells of that organism.

Suddenly the implications of Mendel's work became obvious: chromosomes behaved like the particles or factors that Mendel described. Mendel's hereditary factors were located on the newly discovered chromosomes or were the chromosomes themselves!

Proof, however, that the chromosomes were Mendel's hereditary factors did not come until 1905 when the first physical trait was shown to be the result of the presence of specific chromosomal material and, conversely, that the absence of that specific chromosome meant the absence of the particular physical trait. Microscopic observations had discovered the presence of what have come to be called the **sex chromosomes.** These chromosomes, distinguished from other chromosomes and from each other by their size, were named "**X**" and "**Y**." Researchers in 1905 were surprised to observe that somatic cells taken from female donors always contained two copies of the X chromosome, while somatic cells taken from male donors always contained one copy of the X chromosome and one copy of the Y chromosome. All of the other chromosomes in the nucleated cells of both male and female donors appeared identical. Although scientists were not sure of the mechanism, it seemed quite clear that the sex of an organism was directly related to the identity of the chromosomes in that organism's cells. Thus, sex was shown to be the direct result of a specific combination of chromosomal material, and sex became the first phenotype to be assigned a chromosomal location—specifically the X and Y chromosomes (Figure 1.4).

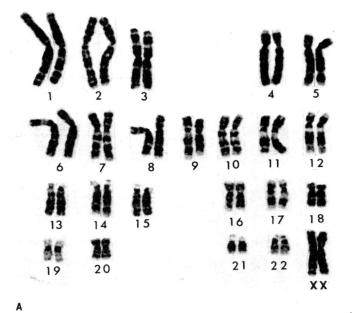

Figure 1.4
Human sex is determined by chromosomes. (a) Chromosomes isolated from the cellular nuclei of normal human females always contain two copies of the X chromosome. (b) In contrast, chromosomes from normal human males always contain one X chromosome and one Y chromosome. All normal human cells—excluding sperm cells, egg cells, and red blood cells—contain 44 additional chromosomes which do not vary with sex.

WHICH CHROMOSOMAL SUBUNIT CARRIES HEREDITARY INFORMATION?

Quantitative analysis of chromosomes shows a composition of about forty percent DNA and sixty percent protein. At first it seemed that protein must be responsible for carrying hereditary information, since not only is protein present in larger quantities than DNA, but protein molecules are composed of twenty different subunits while DNA molecules are composed of only four. It seemed clear that a protein molecule could encode not only more information, but a greater variety of information, because it possessed a substantially larger collection of ingredients with which to work.

This question was finally answered in the early 1950s by using a type of virus called **bacteriophage T2** or **phage T2,** to infect bacterial cells. Scientists Alfred Hershey and Martha Chase carried out experiments in which they prepared one group of phage particles which had incorporated radioactive phosphorus (^{32}P) into their DNA molecules, and another group which had incorporated radioactive sulfur (^{35}S) into the protein molecules of the virus coat. This radioactive labelling allowed phage protein to be distinguished from phage DNA by the different radioactive energies associated with each component.

Hershey and Chase knew that virus particles, including phage T2, function by binding to a target cell and injecting chromosomal material into that cell. The injected viral material contains hereditary information which directs the synthesis of new virus particles using the machinery of the infected cell. Eventually, the cell bursts and releases new virus particles which are able to infect more cells (Figure 1.5). Hershey and Chase asked a most important question: What component of the virus—protein or DNA—directed the synthesis of new virus particles in the infected cell? In other words, does the virus store hereditary information in the viral DNA or in the viral protein? To answer this question, these researchers allowed their radioactively labelled virus particles to infect target cells. After the viral chromosome had been injected into the cell, the cells were agitated to cause the viral coat to break away from the host cell surface. Hershey and Chase then determined the location of the viral DNA and of the viral protein by looking for radioactive phosphorus and radioactive sulfur, respectively. Their experiments detected the presence of only radioactive phosphorus in the infected cells. There was no evidence of radioactive sulfur. These data suggested that because only viral DNA, and not viral protein, could be found in the infected cells, viral DNA must be carrying the hereditary information (Figure 1.6).

The number of experiments which pointed to DNA, and not protein, as the storehouse of hereditary information surprised scientists. Much of this surprise was caused by the fact that the structure of the typical DNA molecule is fairly simple: it consists of only four different subunits that are always joined in the same fashion. It is a long, straight molecule which occurs in a regular spiral shape. Ironically, close analysis of the structure of the DNA molecule put these objections to rest. Let's now make a close examination of the structure of a DNA molecule ourselves.

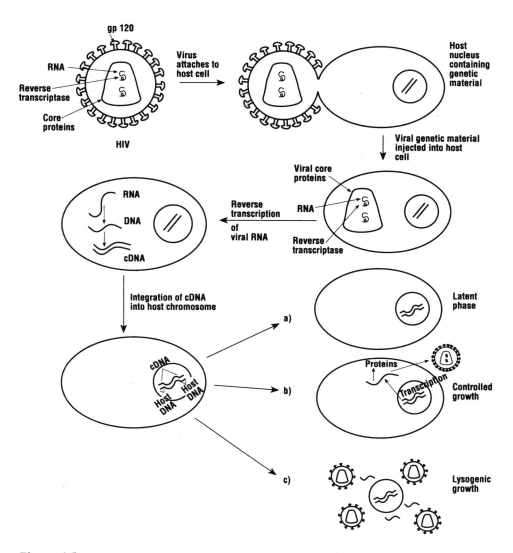

Figure 1.5
An example of viral life cycle. Some specialized virus particles contain RNA, rather than DNA, as their hereditary material. These viruses are known as **retroviruses.** AIDS is caused by a retrovirus known as HIV or **Human Immunodeficiency Virus.** HIV infection begins when the virus attaches to a host cell and injects a combination of enzymes and RNA into the host cell. The enzymes then direct synthesis of a cellular DNA copy of the viral RNA. The DNA copy—known as **cDNA**—then integrates into the host genetic material. The integrated cDNA follows one of three pathways. (a) The integrated cDNA can remain latent and relatively inactive for a time. (b) The cDNA can be used as a template to direct the relatively slow synthesis of new retrovirus particles which will bud from the surface of the host cell or (c) new retrovirus particles can be synthesized so quickly from the cDNA template that the host cell lyses or bursts. Virus particles which contain DNA proceed through stages of infection which resemble that of the retrovirus with the exception of synthesis of a DNA copy, unnecessary because the virus already contains DNA in place of RNA. (From p. 54, *Scientific American,* October 1988. Copyright © 1988 by Scientific American, Inc., George V. Kelvin, all rights reserved.)

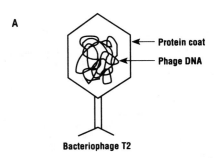

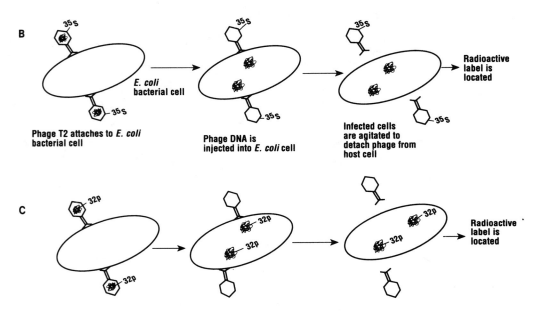

Figure 1.6
The Hershey-Chase experiments demonstrated that genetic information is carried on DNA rather than on protein molecules. (a) The bacteriophage T2 consists of a protein coat which encloses the phage DNA. (b) Scientists labelled the protein coat of phage particles with radioactive sulfur (^{35}S). *E. coli* bacterial cells were then infected with these labelled phage particles. After viral attachment had occurred, the infected cells were agitated to remove any phage material which remained on the cell surface. The ^{35}S-labelled protein was found only in the material that had been separated from the infected cells. No trace of radioactive label was found within the infected cells. (c) The DNA of a second preparation of phage particles was labelled with radioactive phosphorus (^{32}P). As above, these labelled viral particles were allowed to infect *E. coli* host cells. Once again, infected cells were agitated to remove any non-injected viral material. This time, however, radioactively labelled viral DNA could be found within the infected cells. These experiments indicated that while phage protein remained on the outside of the infected cell, phage DNA actually entered the host cell. Therefore, scientists concluded that the hereditary information directing the synthesis of new phage particles by the infected host cell must be contained in the DNA molecules.

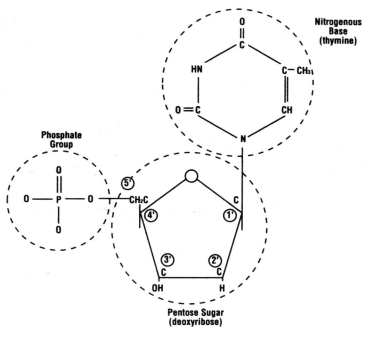

Figure 1.7
DNA molecules are composed of subunits called nucleotides. Each nucleotide is, in turn, composed of three subunits: a pentose sugar containing five carbon molecules which are numbered 1'-5', one of four nitrogenous bases, and a phosphate group.

DNA STRUCTURE

Nucleotides: The Building Blocks of DNA

Everything larger than sub-atomic particles is made of pieces called **subunits.** Some of your subunits are bones, skin, and internal organs; the subunits of a forest include trees and undergrowth; and the subunits of a picket fence are lots of pickets lined up one after the other. A molecule of DNA is also made of subunits: DNA subunits are called **nucleotides.** In fact, even nucleotides themselves are made of subunits. Three types of subunits go into each nucleotide: 1) a sugar molecule called **deoxyribose** which contains five carbon atoms and is therefore called a **pentose sugar,** 2) a **phosphate** group, and 3) one of four **bases,** each of which contains the element nitrogen. These subunits always bond in the same way to make a complete nucleotide (Figure 1.7). Each nucleotide is identified by the particular base which it contains. The four different bases found in DNA nucleotides are called **adenine, thymine, guanine,** and **cytosine** (Figure 1.8).

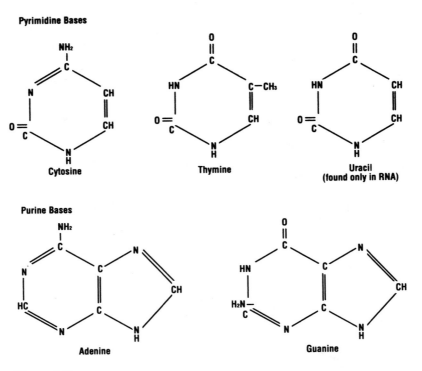

Figure 1.8
The nitrogenous bases of nucleic acids can be divided into two groups. There are three pyrimidine bases: cytosine, thymine, and uracil. Thymine is found only in DNA molecules while uracil is found only in RNA molecules (Chapter 2). There are two purine bases: adenine and guanine.

Nucleotides Are Arranged in Strands

The four nucleotides which make up the vast majority of DNA molecules link together to make a long, straight strand or **polymer**, often called a **polynucleotide strand.** The nucleotides join to one and other by the formation of strong chemical bonds, called **covalent bonds**, between the sugar of one nucleotide and the phosphate group of the next nucleotide (Figure 1.9). All of the bonds holding adjacent nucleotides together are the same: the phosphate group of one nucleotide is attached by a covalent bond to the number 3 carbon (3'-C) of the preceding nucleotide's deoxyribose and the number 5 carbon (5'-C) of the succeeding nucleotide's deoxyribose. This gives the DNA strand **polarity**—that is, one end differs from the other end in the same way that you have a head at one end and feet at the other. A strand of DNA always has a 5'-C at one end and a 3'-C at the other. This bonding pattern which occurs between nucleotides results in a molecule with a regular backbone consisting of alternating sugar and phosphate subunits of adjacent nucleotides. The bases of the nucleotides extend away from the backbone, almost like the steps on a half-ladder.

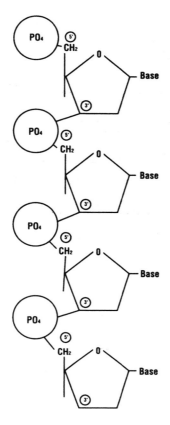

Figure 1.9
Each nucleotide in a nucleic acid molecule is connected to its neighbors via a chemical bond which is formed between the sugar of one nucleotide and the phosphate of the next nucleotide. The resulting polynucleotide molecule ("poly" = many) has a backbone consisting of alternating sugars and phosphate groups. The nitrogenous bases extend from the sugar molecules of the sugar/phosphate backbone.

DNA Molecules Are Double-Stranded

Every DNA molecule is made up of two polynucleotide strands which are twisted around one another to form a structure much like a spiral staircase or ladder (Figure 1.10). The two involved strands run in opposite directions, rather like lanes on a two-way street. If we arbitrarily assign the DNA molecule a top and a bottom, one strand has a 5'-C at the top and a 3'-C at the bottom, while the orientation, or polarity, of the other strand is reversed. Because of this, the two strands of a DNA molecule are referred to as being **antiparallel.**

Both of the polynucleotide strands in a DNA molecule wind around a common central axis to form a spiral shape. The exterior of the DNA spiral is primarily composed of the sugar-phosphate backbones of the two polynucleotide strands. The nucleotide bases extend inward toward the central axis. The two polynucleotide strands are held together by relatively weak chemical bonds, called **hydrogen bonds,** which occur between the bases of each strand. The overall structure of the DNA molecule can be thought of as resembling a spiral staircase in which the handrailings of the staircase represent the sugar-phosphate backbones of the two strands, and each stair represents a pair of bases held together by hydrogen bonds. This spiral structure of DNA is called a **double helix** and was proposed by James Watson and Francis Crick working at Cambridge University in England in the early 1950s. Many scientists have since carried out research whose results both confirm and further elucidate the structure of the DNA molecule.

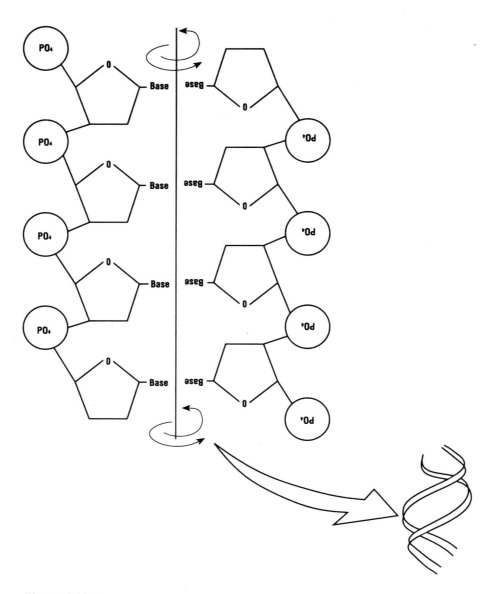

Figure 1.10
A DNA molecule consists of two polynucleotide strands twisted about one another to form a spiral structure. The resulting molecule contains two anti-parallel exterior sugar/phosphate backbones. The interior of the molecule contains the base pairs whose hydrogen bonds maintain the integrity of the double-stranded DNA molecule.

The Pairing of Bases in the Double Helix Is Not Random

The bases which are found in a DNA molecule can be divided into two categories based on their chemical structure. Thymine and cytosine are **pyrimidines,** which have a single ring structure, while the **purines,** guanine and adenine, have a double ring structure (Figure 1.8).

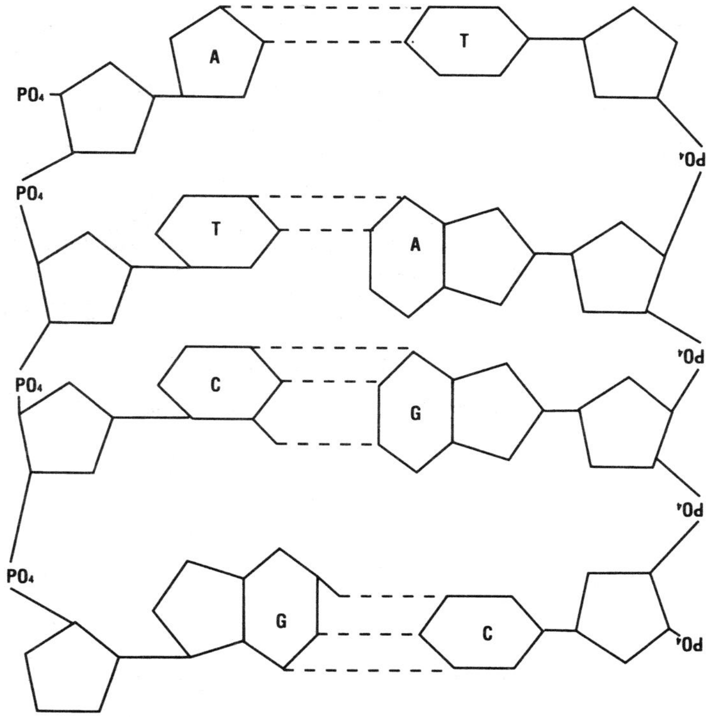

Figure 1.11
Only four types of base pairing are possible in a DNA molecule. Adenine and thymine are held together by two hydrogen bonds, while guanine and cytosine are held together by three.

The hydrogen bonds which hold together the two strands of a DNA molecule do not occur at random along the length of that molecule. Instead, they always occur between one purine and one pyrimidine base. In this way the diameter of the double helix maintains a constant 2.0 nanometers (1 nanometer = 1/1,000,000th of a millimeter) along its entire length. More specifically, thymine always bonds to adenine by two hydrogen bonds, while guanine always bonds to cytosine by three hydrogen bonds. This is called complementary base pairing. Each base pair may appear as a rung on the DNA ladder in either of two possible orientations. Thus, there are four possible DNA rungs or complementary base pairings: A:T, T:A, G:C, and C:G (Figure 1.11).

These restrictions on base pairing have a most important result. Any time the identity of one base in a pair is known, the other can automatically be deduced. For example, if one base in a pair is T, the other must be A. If one base is G, the other must be C. Perhaps even more important, if the sequence of bases is known along one strand of a DNA molecule, it is a relatively simple matter to deduce the sequence of the other strand: if one strand is experimentally found to have the sequence AATCGTCG, then the other strand *must* have the sequence TTAGCAGC. These matching sequences are described by the term **complementary**, and a double-stranded DNA molecule is said to consist of **complementary strands**.

DNA REPLICATION

The DNA Molecule Must Copy Itself Accurately

The processes of development, growth, and repair of an organism involve cell division. For example, the repeated divisions of a newly fertilized, single-celled zygote allow the development of a newborn human baby made up of trillions of cells. We know that the complete process of cell division requires the chromosomal material inside the cell to double just prior to the actual splitting of the cell. The doubled chromosomal material is divided equally between the two new cells. Analysis of the new cells shows that both sets of chromosomes are identical to one another. How does the DNA molecule make copies of itself enough times to provide all of the trillions of cells of that newborn baby with the right DNA? Perhaps even more important, how does the genetic information contained in the DNA molecule reproduce with enough accuracy to result in a normal newborn in the vast majority of births? We know that the bottom copy from a stack of carbon papers is always less clear than the top copy. Why don't multiple copies of DNA suffer the same fate?

DNA Complementarity Allows Replication to Occur

The answer to these questions lies in the complementary structure of the DNA molecule. As we discussed above, knowledge of the base sequence on one DNA strand allows us to predict, with enormous accuracy, the base sequence of the other strand. If one strand has the base sequence AATTGCG, then the other strand *must* read TTAACGC. This complementarity means that each strand of a DNA molecule can act as a **template,** or pattern, for the production of its own complementary strand. Let's see how this fact plays a role in **replication**—the production of new DNA molecules.

DNA Replication Is Semi-Conservative

In 1958, scientists Matthew Meselson and Franklin Stahl, working at the California Institute of Technology, showed that DNA replication operates in a **semiconservative** fashion. This means that when one DNA molecule undergoes replication to produce two identical new molecules, each of the new molecules is made up of one polynucleotide strand from the original molecule bonded to one newly synthesized polynucleotide strand. To demonstrate this fact, these scientists grew a type of bacteria known as ***Escherichia coli*** in the presence of nutritional media containing "heavy" nitrogen, or ^{15}N. As a result, the bases of the *E. coli* bacterial DNA contained heavy nitrogen in place of the more common, lighter, ^{14}N. In other words, the bacterial cells contained "heavy" DNA—so called because it had incorporated heavy nitrogen into its nucleotide bases. Heavy DNA could be distinguished from the normal, or light, DNA which could be isolated from cells grown in the presence of ^{14}N. This distinction was made by determining the relative molecular densities of the two types of DNA molecules.

Meselson and Stahl wanted to know what kind of DNA molecules—heavy, intermediate, or light—would be found if cells grown only with ^{15}N were allowed to undergo a single round of replication in the presence of ^{14}N. To answer this question, the scientists placed ^{15}N-treated cells in a growth liquid which contained ^{14}N as the only source of nitrogen. They then isolated DNA from the bacterial cells after a single round of replication. Analyses showed that those cells which had replicated contained only DNA molecules whose molecular weights were intermediate—that is, lighter than ^{15}N-DNA and heavier than ^{14}N-DNA. DNA isolated after a second round of replication contained both molecules of intermediate density and molecules of light density, but no heavy DNA molecules. These results could only be explained by a semi-conservative mode of replication. In other words, after the first round of replication, all DNA molecules were made up of one heavy and one light strand, thus yielding molecules of intermediate weight. After a second round of replication, one half of the DNA molecules still contained one heavy and one light strand while the remaining molecules contained two light strands (Figure 1.12). Thus, in DNA replication each strand acts as a template for the synthesis of a new complementary strand, and each new DNA molecule is made up of one newly synthesized strand and one strand passed down from the parent molecule.

DNA Replication Has Three Stages

As we have discussed above, the complementary nature of the DNA molecule means that each of the two strands in the molecule carries the information necessary to produce a new complementary strand. This complementarity is absolutely necessary for DNA replication to occur, since the base sequence of one DNA strand automatically specifies the base sequence of the other. Although not all of the details surrounding the process of replication have been elucidated as of the preparation of this text (1991), DNA replication can be described in three basic steps:

1. **Unwinding**—the two strands of the DNA double helix molecule must unwind and separate from one another in the areas which are being actively replicated. This occurs when a series of **enzymes** break the weak hydrogen bonds holding together the base pairs which form the stair steps of the double helix. (An enzyme is a type of molecule that helps a particular chemical reaction to take place.) Unwinding a segment of a double-stranded DNA molecule results in two lengths of unpaired single strands of nucleotides. Each single-stranded length has "free" bases attached to its sugar-phosphate backbone.

2. **Complementary base pairing**—free nucleotides containing various unpaired bases are always present in the interior of living cells. The bases of these nucleotides attach by the formation of hydrogen bonds to complementary bases on each of the single strands generated by the unwinding process. In this way, the order of the bases on the newly forming strands is directed by, and is complementary to, the parent strand.

3. **Polymerization**—an enzyme called **DNA polymerase** creates the backbone of the newly synthesized nucleotide strand by forming covalent bonds which connect the sugar subunit of one nucleotide to the phosphate subunit of the adjacent nucleotide in the growing DNA strand.

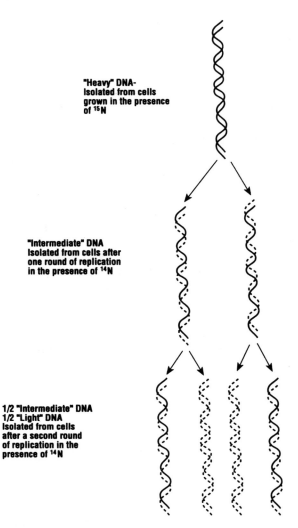

Figure 1.12
The Meselson-Stahl experiments showed that DNA replication is a semi-conservative process. "Heavy" DNA grown in the presence of ^{15}N can be separated from "light" DNA grown in the presence of ^{14}N. Bacteria were grown in ^{15}N-growth medium until all bacterial cells contained only heavy DNA molecules. These cells were then transferred to ^{14}N-growth medium and allowed to undergo a single round of DNA replication. DNA isolated from these cells showed a molecular weight which was intermediate—heavier than "light" ^{14}-DNA but lighter than "heavy" ^{15}N-DNA. Cells allowed a second round of DNA replication in ^{14}N-growth medium yielded half "intermediate" DNA molecules and half "light" DNA molecules. These results are consistent with a semi-conservative mode of DNA replication in which each new DNA molecule contains one parental strand and one newly synthesized strand.

When all of the bases on the separated strands of parental DNA are matched with their complementary bases, two new daughter DNA molecules result. Each of the new daughter molecules is identical in base sequence to the original DNA molecule, since each is composed of one parental nucleotide strand and one newly synthesized complementary strand.

DNA Replication Has a Built-In Correction Mechanism

Estimates suggest that a mistake occurs in the complementary base pairing of DNA replication only about three times during the replication of all of the DNA in a germ cell. That may not seem like much—three mistakes on a single typewritten page is often cause for celebration. But consider this: only about 3800 letters appear on the average typewritten page, while there are approximately 3×10^9 "letters" or base pairs in the average germ cell! To get the same level of accuracy in your typing as there is in DNA replication you would have to type approximately 79,000 pages with only three mistakes! How is this phenomenal accuracy achieved? When you read each page after you have typewritten it, you look for mistakes—you proofread. Special enzymes carry out much the same function during the process of replication. These enzymes "read" the nucleotides in the newly synthesized DNA strand. Any time they come to a base that is not complementary to the base on the template strand, they **excise,** or remove, the problematic base. Soon the correct base hydrogen bonds to the template, thus correcting the base sequence and making it a perfect match for the original parent molecule. If an incorrect base is not removed from the new DNA molecule, it results in a **mutation** or change in hereditary information. Mutations are discussed in depth in chapter three.

REVIEW AND SUMMARY

1. Gregor Mendel conducted a series of quantitative genetic experiments whose results led him to postulate that all phenotypes are the result of the interaction of two "factors" which carry hereditary information.

2. Experimental evidence showed that nuclear chromosomes behaved in a way that matched the requirements which Mendel had predicted for his factors. Scientists concluded that hereditary information must reside on the chromosome, composed of both protein and DNA.

3. Although initially it seemed that hereditary information must reside on the protein component of a chromosome, various experiments proved differently. Instead, we now know that hereditary information is encoded within the relatively simple structure of the DNA molecule.

4. DNA is a linear molecule composed of subunits called nucleotides. Nucleotides, in turn, are composed of three types of subunits: a 5-carbon sugar called deoxyribose, a phosphate group, and a nitrogenous base.

5. A DNA molecule is composed of two polar chains of nucleotides which are twisted around a common axis, thus forming a double helix structure. Each nucleotide chain has a backbone consisting of repeating sugar-phosphate subunits which are held together by relatively strong covalent bonds. The nitrogenous bases extend at right angles away from the backbone and toward the central axis of the molecule.

6. Relatively weak hydrogen bonds, which join pairs of bases, hold the two chains together. Base pairing occurs in a specific fashion in which two possible pairs, each of which can occur in two orientations, are possible. Thus, the four base pairs that occur are A:T, T:A, G:C, and C:G. These base pair restrictions result in the complementarity of the DNA molecule.

7. The three steps of DNA replication include: 1) unwinding of the double helix, 2) complementary pairing of the nucleotide bases on the newly single strands with free bases in the cell, and 3) polymerization of the backbone on each newly synthesized strand.

8. Complementarity is at the heart of DNA replication. Each strand of a DNA molecule carries all of the information necessary for the formation of a new complementary strand, and thus for the formation of two new complete DNA molecules where one existed previously.

9. DNA replication is extraordinarily accurate, with approximately three mistakes occurring for every 3×10^9 base pairs completed. Mistakes in base pairing are most often corrected by "proofreading" enzymes in the cell itself.

DNA: The messenger of life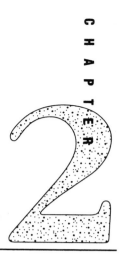

When a sperm fertilizes an egg (Figure 2.1) it results in the combination of two sets of hereditary information, one from the mother and one from the father, to create a new organism which is unlike anything ever created before. The system usually works beautifully and a normal, healthy baby (Figure 2.2) results. Sometimes, however, something goes awry.

Jamie S., the newborn son of Darlene and Jon, grew and developed normally for the first eighteen months of his life. But when Jamie was two years old it became obvious that he was not like other babies. Jamie was lethargic, he tired easily, and he seemed to have trouble breathing. Thinking that Jamie might have asthma, Darlene and Jon took him to the doctor. But when the doctor examined Jamie's red blood cells he saw that they were misshapen—instead of being relatively smooth and biconcave, they were sickle-shaped (Figure 2.3). Jamie had a disease called sickle-cell anemia.

Jamie's disease was the result of a small mistake—a **mutation**—in the genetic material of one of his chromosomes (Figure 2.4). This mutation, in which one of approximately 1700 of the involved base pairs is "wrong," meant that Jamie's body could not correctly produce a complex molecule called **hemoglobin.** A protein, hemoglobin is composed of four **polypeptides** or chains of amino acids, and four **heme** groups. Each red blood cell contains approximately 2.8×10^8 molecules of hemoglobin; the heme groups function by binding to oxygen molecules in the lungs, and later releasing those oxygen molecules into the body's cells and tissues. Red blood cells which contain the sickle cell hemoglobin take on the characteristic bent shape of a sickle, hence the name of the disease. The body's immune system often destroys affected red blood cells, thus leading to severe anemia and to a marked decrease in the amount of oxygen available to the body. In addition to tiring easily, sickle-cell anemia patients may experience abdominal and joint pains, kidney and liver damage, and a host of other symptoms.

24 DNA: The Messenger of Life

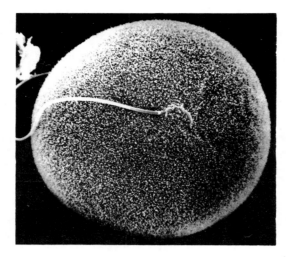

Figure 2.1
Human sperm and egg co-mingle their genetic material at fertilization, thus creating a genetically unique individual.

Figure 2.2
It is amazing to think that a single fertilized egg contains all of the detailed genetic instructions needed to create a complex and unique new person. Above, 2 week old Emilie is admired by her brothers and sister.

How can such a small mistake—one single base pair out of approximately 1700 in the DNA sequences that encode the hemoglobin polypeptides—result in changes in a protein that are massive enough to convert a normally smooth, biconcave red blood cell into the sharp, irregular, pointed shape characteristic of a red blood cell in sickle-cell anemia? How can the substitution of a *single* base pair in the hemoglobin gene cause the devastating condition of sickle-cell anemia? To understand how this occurs we must understand how a segment of DNA directs the production of a protein molecule.

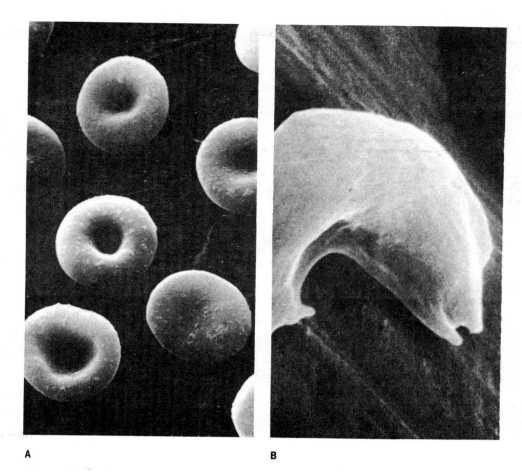

A **B**

Figure 2.3
(a) Normal red blood cells exhibiting typical smooth, biconcave shape × 5970.
(b) Sickled red blood cells isolated from a person with sickle-cell anemia. Note the characteristic bent sickle shape of these abnormal red blood cells × 19,000.

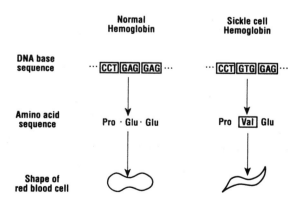

Figure 2.4
Sickle-cell anemia results from a single change in the base sequence of the gene that encodes the hemoglobin protein. That change leads to an amino acid substitution in the mature hemoglobin molecule: valine takes the place of glutamic acid at a single location. As a result of this relatively minor molecular change, the red blood cell takes on the sickle shape characteristic of sickle-cell anemia.

OVERVIEW In chapter one we learned that DNA molecules are responsible for the passage of phenotypic traits from generation to generation. This chapter explains how this is possible. The DNA in chromosomes is divided into functional subunits called **genes,** and each gene holds the blueprint for a particular polypeptide product. We will see how the information encoded in the gene is transferred to the polypeptide molecule it produces, which, in turn, helps to define the phenotype of an organism. This chapter shows us how changes in the base sequence of a gene can change a final phenotype and answers the following questions:

1. What is the main function of a DNA molecule? What is the relationship of structure to function in the DNA molecule?

2. What are genes and what is their language?

3. How does the hereditary information flow from the nucleus, where the genes are located, to the cytoplasm, and what are the major players in the process?

THE RELATIONSHIP OF GENES TO PHENOTYPES

One Gene–One Polypeptide

You remember that the experiments of Gregor Mendel suggested that phenotypic traits resulted from interactions between two factors which we now call **genes,** and that later work by other scientists showed that these factors must reside on the DNA component of the chromosome. What exactly is the relationship of a gene to a phenotype?

In the early 1940s scientists George Beadle and Edward Tatum at Stanford University studied the biochemistry and enzymology of a mold known as **Neurospora.** These researchers found that when *Neurospora* cells were exposed to ultraviolet light and x-rays, known causes of hereditary changes, some of the *Neurospora* colonies lost their ability to utilize various nutrients in their growth medium. Each mutant colony required various amino acids, vitamins, or other growth factors to supplement its normal diet. Further research showed that, in each case, the mutant colonies had lost the ability to synthesize a particular **enzyme.** The missing enzyme would have converted one or more of the nutrients from the *Neurospora's* growth medium into the particular molecule with which the colony now had to be supplemented. In other words, a normal strain of *Neurospora* might use enzyme X to convert nutrient A into product B. A mutant strain, lacking enzyme X, would be unable to complete the conversion of A to B, and would thus require a direct source of B for continued survival (Figure 2.5). These data suggested to the researchers that a single gene was responsible for the production of a single enzyme.

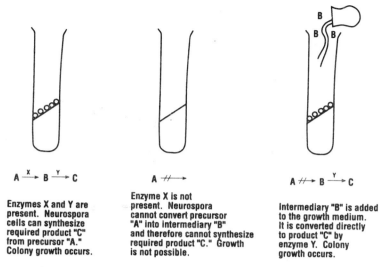

Figure 2.5
Growth of *Neurospora* requires either that all necessary nutrients be supplied in the growth medium, or that the *Neurospora* cells themselves possess specific enzymes able to convert supplied substrates into the required nutrients. If those enzymes are not present, necessary nutrients will not be synthesized, and colony growth will be slowed, or even halted entirely.

The conclusions of Beadle and Tatum were expanded in 1956 when scientist Vernon Ingram demonstrated that at least some of the molecular abnormalities of the hemoglobin protein in sickle-cell anemia resulted from a single change in one of the genes for hemoglobin. Thus in the work of both groups of researchers, a mutation in a gene resulted in a change in a long chain of amino acids called a polypeptide. In the Beadle and Tatum experiment the polypeptide was an enzyme, while in the Ingram experiment the polypeptide was a protein without enzymatic function. However, despite the functional difference, both molecules were polypeptides. Furthermore, specific phenotypic changes—novel nutritional requirements in one case and abnormally formed hemoglobin molecules in the other—could be directly related to the change in, or absence of, a polypeptide. These important experiments provided the first proof that a change in a gene could lead to a change in a polypeptide and, from there, to a change in phenotype.

These pieces of research were combined as evidence for the concept that each gene on a chromosome produces a single type of polypeptide, summarized by the phrase "one gene–one polypeptide." This phrase is sometimes stated as "one gene–one protein" because one or more long chains of amino acids make up a protein. Proteins and polypeptides differ, however, in that proteins are sometimes made up of multiple polypeptide chains. In these instances, including the hemoglobin protein which is made up of four polypeptides, the individual polypeptide chains are not functional proteins. In many instances, however, the terms "polypeptide" and "protein" are used interchangeably.

HEREDITARY INFORMATION TRAVELS IN ONE DIRECTION ONLY

Imagine a cookbook large enough to hold recipes for every type of dish ever prepared in the United States—a cookbook so large that it would be necessary to store it on a special stand, perhaps in your living room or den. If you decided, one day, to make a nice batch of chocolate chip cookies, you would go to the cookbook and open it to the correct page. Then, rather than cart the enormous book into the kitchen, you would copy the cookie recipe onto a 3" × 5" index card, take the card into your kitchen, and put it onto the counter. You would read the card and, one by one, collect all of the ingredients necessary to prepare the cookies. Once all of the ingredients were assembled at the counter you would mix them together, in a specific order, to make the cookie dough. Then you would bake the dough and eat the cookies.

Although you may consider this story just something to make your mouth water, you could also consider it as an analogy to DNA function and protein synthesis. The cookbook represents the DNA in the nucleus of each one of your cells. The DNA in each cell contains all of the information or "recipes" necessary to construct all of the different polypeptides that make up your body. Each individual recipe represents a gene. When the information in a single gene is needed, a copy of the information is made in the form of a molecule called **messenger RNA** or **mRNA**. The mRNA functions like the 3" × 5" index card. Like the card, the mRNA is carried to the work site, in this case not a kitchen counter, but rather the **ribosomes** found in the cell's cytoplasm. Once the mRNA is placed on the worksite, or ribosome, other special molecules called **transfer RNA,** or **tRNA,** collect all of the ingredients needed to make the polypeptide which is specified by the recipe. In this case, of course, the ingredients are amino acids since polypeptides are chains of amino acids. The tRNA molecules bring the amino acids back to the ribosome and, following directions contained in the mRNA, put them together to make the polypeptide product.

DNA to RNA to Polypeptide: The Central Dogma

This series of events in which information is transferred from the DNA to the RNA, and from there to the final polypeptide product, is called the **Central Dogma** of genetics. The Central Dogma refers to the fact that hereditary information normally travels in one direction only: from DNA to RNA to protein. Each step of the information transfer process has a special name. When information is copied from DNA to RNA the process is called **transcription,** and when information is copied from RNA to polypeptide the process is called **translation** (Figure 2.6).

BOTH RNA AND DNA ARE NUCLEIC ACIDS

Both RNA and DNA molecules are nucleic acids and, as such, have many structural similarities. Both are long, linear molecules composed of nucleotides. A phosphate group, a sugar, and a nitrogenous base compose both the ribo- and deoxyribonucleotides (Figure 1.8). The successive nucleotides in each molecule are joined together with covalent bonds between the phosphate of one nu-

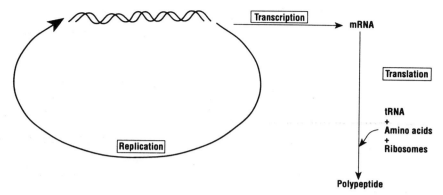

Figure 2.6
The Central Dogma refers to the fact that hereditary information normally travels in one direction only: information contained in DNA is copied either into new molecules of DNA in the process of replication, or into RNA in the process of transcription. From RNA, information is converted into the final polypeptide product in the process of translation.

cleotide and the sugar of the next. This results in a sugar-phosphate backbone in both the RNA and DNA molecules. But the major similarities end there. Listed below are some of the major differences:

1. Ribonucleotides and deoxyribonucleotides differ in the bases that they contain. The DNA bases are adenine, thymine, guanine, and cytosine. Although the RNA bases include adenine, guanine and cytosine, **uracil** takes the place of thymine. Not only does uracil resemble thymine structurally, but it is also complementary to adenine (Figure 1.8).

2. The sugar molecule in ribonucleotides is **ribose,** while the sugar in deoxyribonucleotides is deoxyribose.

3. RNA is a single-stranded molecule, while DNA consists of two complementary polynucleotide strands.

4. Chromosomal DNA stays in the nucleus of the non-dividing cell at all times, while RNA is transported into the cytoplasm to carry out its functions.

5. RNA molecules are generally much shorter than DNA molecules. In addition, the life span of an RNA molecule is much less than that of a DNA molecule, which remains active for the entire lifespan of the cell.

6. Only one major type of DNA molecule actively participates in the life processes of the cell while there are four distinct types of RNA. **Messenger RNA** (mRNA) contains the hereditary information copied from the DNA; **ribosomal RNA** (rRNA) is a major component of the ribosomes which are the cytoplasmic sites of protein synthesis; **transfer RNA** (tRNA) functions to carry amino acids to the ribosome and fits them into the proper position as specified by the mRNA; and **small nuclear RNA** (snRNA) is involved in the conversion of immature to mature mRNA molecules.

PRODUCTION OF RNA

Transcription Takes Place in Four Basic Steps

Transcription is the process whereby the hereditary information contained in a gene is transferred onto a newly synthesized strand of RNA called mRNA, because it contains the gene's message. Although DNA, and therefore genes, are double stranded, only a single strand of the gene contains the information needed to synthesize the polypeptide specified by the gene. This template strand is called the **coding strand.** The other strand, which is complementary to the coding strand, is called the **non-coding** or **anti-sense strand.**

Transcription takes place in four steps:

1. **Unwinding**—the DNA double helix must unwind in order to free the coding strand of the gene which is being transcribed.

2. **Complementary base pairing**—free nucleotides, which contain the RNA bases (A, U, G, or C), must pair with the exposed bases of the deoxyribonucleotides on the coding strand of DNA.

3. **Elongation**—an enzyme called **RNA polymerase** must join the ribonucleotides in the correct order to form a new RNA molecule.

4. **Separation**—the new RNA molecule separates from the DNA template and the DNA double helix re-forms.

Let's look at these steps individually and in more detail (Figure 2.7).

STEP ONE: THE DNA UNWINDS TO FREE THE GENETIC TEMPLATES In order for the information contained in a gene to be copied into a molecule of RNA, the DNA double helix must unravel in the immediate area surrounding the gene. The unwinding of the DNA double helix is aided or **catalyzed** by the binding of an enzyme to the coding strand of the DNA molecule. This enzyme, called RNA polymerase, binds just **upstream,** or toward the 5′ end, of the gene in a sequence of bases known as the **promoter.** The promoter sequence functions not only as a binding site for the RNA polymerase, but also serves to indicate the correct beginning of the gene which is being transcribed.

Examination of a number of cell types has shown that the base sequences of many promoters are remarkably similar. For instance, **prokaryotic cells**—cells without nuclei—such as *E. coli* bacterial cells, have the base sequence TATAAT immediately upstream of most genes, while many **eukaryotic cells**—cells with nuclei—have the base sequence TATA. These sequences, called the **Pribnow box** and the **TATA box,** respectively (see chapter eight), act to promote transcription by signaling a polymerase enzyme to bind to the DNA double helix in preparation for the transcription of a nearby gene.

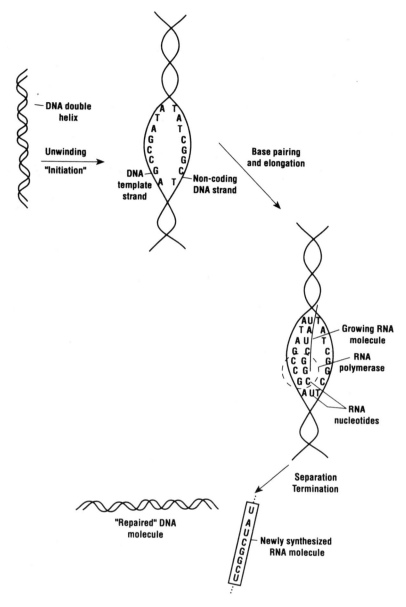

Figure 2.7
The transcription of DNA into RNA takes place in four basic steps. **Initiation** involves the unwinding of the DNA double helix in the area surrounding the DNA sequence to be transcribed. Free ribonucleotides then bond to the exposed complementary bases on the coding strand of the DNA **(base pairing)**. **Elongation** involves the bonding of the correctly ordered ribonucleotides to one another in the growing RNA molecule. As soon as the first two RNA nucleotides are correctly positioned, the enzyme RNA polymerase joins the sugar of one to the phosphate of the next, eventually creating the sugar/phosphate backbone of the RNA molecule. **Separation** or **termination** involves the separation of the completed RNA molecule from the DNA template, and the re-bonding of the two complementary DNA strands.

STEP TWO: COMPLEMENTARY BASES PAIR WITH FREE CODING BASES Once the DNA molecule unwinds, bases on the coding strand become "free" or unpaired much as occurs in the process of DNA replication. Free ribonucleotides, the subunits of RNA molecules, are present within the cell and come in contact with the unpaired bases of the coding strand. If the bases of the free ribonucleotides and the deoxyribonucleotides of the coding strand are complementary, hydrogen bonds form to hold each pair together temporarily. The anti-sense, or non-coding, DNA strand is not involved in this complementary base pairing process.

STEP THREE: RIBONUCLEOTIDES ARE JOINED BY COVALENT BONDS As the bases of the gene's coding strand pair with complementary RNA bases, cellular enzymes join adjacent ribonucleotides to one another by the formation of covalent bonds between the phosphate group of one ribonucleotide and the sugar group of the adjacent ribonucleotide. This forms a backbone of alternating sugar-phosphate subunits, as also occurs in the DNA molecule.

STEP FOUR: SEPARATION OF THE NEW RNA MOLECULE Once the bases of the entire gene pair to complementary bases in ribonucleotides, and the ribonucleotides join by covalent bonds, the DNA template strand releases the completed new RNA molecule. The RNA polymerase is also released at this time. The two strands of the DNA molecule then finish rejoining by formation of hydrogen bonds between their complementary bases. Its transcriptional activities leave the DNA molecule essentially unchanged.

THE RNA MOLECULE: STRUCTURE AND FUNCTION

RNA Directly Complements DNA

A most important result of the transcription process is that an RNA molecule has been synthesized which is complementary to the DNA template which directed its synthesis. In other words, if we know the identity of the bases in a sequence of DNA, we can predict the identity of the bases in the corresponding sequence of RNA. If even a single base is changed in the DNA molecule, then a corresponding base will be changed in the RNA molecule. The complementarity of sequences in transcription resembles the complementarity that occurs between the DNA strands in the process of replication. Just as we can predict the sequence of one DNA strand by examining the base sequence of the other, we can also predict the sequence of an RNA molecule by examining the gene responsible for its formation.

RNA Is Often Processed Prior to Translation

Once a newly transcribed mRNA molecule is released from its DNA template, it often is processed prior to its use as a template in protein production. The first processing step is the addition of a long stretch of adenine ribonucleotides to the 3' end of the mRNA molecule. This sequence of newly added ribonucleotides is called a **poly(A) tail,** and the process by which it is added is called **polyadenylation.** In addition, the 5' end of the mRNA may be attached to a cap structure consisting of a modified guanine base, which is joined to the mRNA molecule with a reverse orientation. In other words, while most ribonucleotides are joined to one another via covalent bonds between their 5' and 3' carbons, the 5' cap is added to an RNA molecule so that its 5' carbon bonds to the 5' carbon of the adjacent ribonucleotide. This processing—the addition of a 5' cap and a 3' poly(A) tail—generally occurs only in the RNA of eukaryotic cells and appears to protect the mRNA molecule from premature degradation.

Newly produced, or **nascent,** eukaryotic mRNA is often longer than mature, functional mRNA due to the presence of spacers known as "intervening sequences" or **introns,** in the eukaryotic template gene. The base sequences present in the introns do not contain hereditary information. Hereditary information, converted into a polypeptide product, is, instead, held in base sequences known as **exons**—sequences which contain the coding regions of a gene. A series of alternating introns and exons, both of which are transcribed into the complementary RNA molecule, compose most eukaryotic genes. In order for hereditary information to be converted from the mRNA molecule into a polypeptide product, the non-coding introns must be removed from the immature mRNA molecule and the remaining exons must be joined together or **ligated.** This processing step is known as **intron splicing** (Figure 2.8).

Two Types of RNA Do Not Carry Hereditary Information

As discussed above, not all RNA molecules carry messages to direct the synthesis of a polypeptide product. The two other types of RNA—rRNA and tRNA—are discussed below. Ribosomal RNA is used in the formation of ribosomes—cytoplasmic organelles which are the sites of protein synthesis. Ribosomal RNA, unlike mRNA, is not ligated after the removal of introns. This results in the production of multiple rRNA molecules from a single unprocessed precursor. There are four types of rRNA molecules, each with a characteristic molecular length and weight, which become associated with approximately seventy-five different protein molecules to form a single ribosome.

Transfer RNA, or tRNA, is the RNA which reads the hereditary information contained in the mRNA molecule. A molecule of tRNA is about 75 ribonucleotides in length and is folded into a modified cloverleaf shape which is, in turn, folded onto itself. This complex structure of a folded cloverleaf, characteristic of all tRNA molecules, occurs because of chemical attractions between some of the tRNA ribonucleotides.

Transfer RNA molecules function much as language translators do. If you speak two languages you may be called on to listen to a sentence in one language, and then to convert that sentence into words of a second language. A tRNA molecule works in a similar fashion; its first language is "RNA" and its second language is "amino acid." Once the tRNA has translated the information from "RNA" to "amino acid," it then transfers the appropriate amino acids from the cytoplasm to the new polypeptide chain which is being constructed on the ribosome.

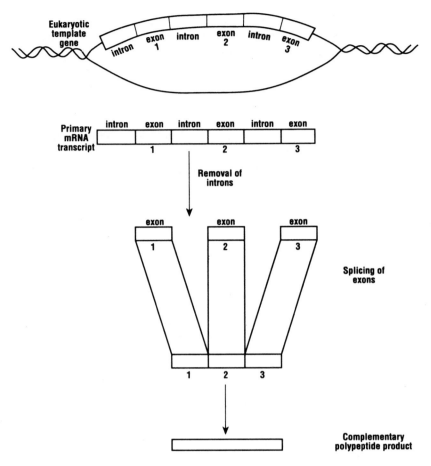

Figure 2.8
Eukaryotic genes are composed of both introns and exons. Although both introns and exons are transcribed into RNA, only exons contain genetic coding information. Thus, the non-coding introns must be removed from the immature RNA molecule, and the remaining exons spliced together to form the mature RNA molecule. The spliced mature RNA molecule, in turn, serves as the template for translation of the functional polypeptide product.

The Language of RNA: The Genetic Code

In order to follow your teacher's directions, you must listen to what the teacher is telling you and then process the teacher's message into something that you understand. If, for instance, your first language is English and you are taking a Spanish class, the teacher will probably address you in Spanish. Until you are rather fluent in Spanish, you will probably translate the teacher's words into English for better comprehension. Similarly, a molecule of tRNA translates information from "RNA" into "amino acid": if the tRNA is to follow the directions of the mRNA and construct a new polypeptide, it must understand the words of the "RNA" language. Those words are called the **genetic code.**

First base	Second base				Third base
	U	*C*	*A*	*G*	
U	UUU ⎱ Phenylalanine UUC ⎰ UUA ⎱ Leucine UUG ⎰	UCU ⎫ UCC ⎬ Serine UCA ⎪ UCG ⎭	UAU ⎱ Tyrosine UAC ⎰ UAA† ⎱ Stop UAG† ⎰	UGU ⎱ Cysteine UGC ⎰ UGA† Stop UGG Tryptophan	U C A G
C	CUU ⎫ CUC ⎬ Leucine CUA ⎪ CUG ⎭	CCU ⎫ CCC ⎬ Proline CCA ⎪ CCG ⎭	CAU ⎱ Histidine CAC ⎰ CAA ⎱ Glutamine CAG ⎰	CGU ⎫ CGC ⎬ Arginine CGA ⎪ CGG ⎭	U C A G
A	AUU ⎱ Isoleucine AUC ⎰ AUA AUG* Methionine	ACU ⎫ ACC ⎬ Threonine ACA ⎪ ACG ⎭	AAU ⎱ Asparagine AAC ⎰ AAA ⎱ Lysine AAG ⎰	AGU ⎱ Serine AGC ⎰ AGA ⎱ Arginine AGG ⎰	U C A G
G	GUU ⎫ GUC ⎬ Valine GUA ⎪ GUG ⎭	GCU ⎫ GCC ⎬ Alanine GCA ⎪ GCG ⎭	GAU ⎱ Aspartic Acid GAC ⎰ GAA ⎱ Glutamic Acid GAG ⎰	GGU ⎫ GGC ⎬ Glycine GGA ⎪ GGG ⎭	U C A G

*Initiation codons. The methionine codon AUG is the most common starting point for translation of a genetic message.
†Terminator codon.

Figure 2.9
The genetic code represents the dictionary of genetic codons and their amino acid translations. (From Linda Maxson and Charles Daugherty, *Genetics: A Human Perspective,* 2d ed. Copyright © 1989 Wm. C. Brown Publishers, Dubuque, Iowa. All Rights Reserved. Reprinted by permission.)

The alphabet of the genetic code is simple; it consists of only four "letters." These letters are the bases found in ribonucleotides—adenine, uracil, guanine, and cytosine or A, U, G, and C. Unlike most languages which have words of varying lengths, all of the words in the genetic code are made up of three letters, each one of which represents one of the four ribonucleotide bases. The genetic code contains only sixty-four different words, because there are only sixty-four possible combinations of four letters into three-letter words. Each word, called a **codon,** specifies one of the twenty amino acids which make up polypeptides. By constructing mRNA molecules with known sequences of ribonucleotides, scientists were able to construct a dictionary defining which amino acid is specified by each of the sixty-four codons. When these artificial mRNA molecules were placed in an environment that contained everything needed for protein synthesis, they produced actual polypeptides. For example, when scientists made an mRNA molecule with only adenine ribonucleotides (AAAAAAAA), it translated into a polypeptide made of only the amino acid lysine (lys-lys-lys). When scientists made an mRNA out of only uracil ribonucleotides (UUUUUUUU), it translated into a polypeptide made of only the amino acid phenylalanine (phe-phe-phe). Constructing all possible triplet combinations of the four ribonucleotides made it possible to generate a complete dictionary of the genetic code (Figure 2.9).

By looking at Figure 2.9, you can see that some amino acids are specified by more than one codon. For example two codons—UUU and UUC—specify phenylalanine, while four codons—GUU, GUC, GUA, and GUG—specify valine. No codon, however, specifies more than one amino acid. This means that although there are synonyms in the genetic code there is no ambiguity: any single codon can have only one specific meaning.

In most languages certain symbols indicate the beginning and the end of a sentence. In English, a capital letter signals the beginning, while the end is signalled by a punctuation mark. In the language of the genetic code, the presence of a **start codon** sets apart the beginning of a sentence or a gene. This codon—AUG—specifies the amino acid methionine and is usually used to indicate the place at which translation should start. One of three STOP CODONS marks the end of the gene. Wherever one of these codons—UAA, UAG, or UGA—appears, translation stops. These four codons act as the punctuation for translation.

How Does tRNA Read the Genetic Code?

Transfer RNA molecules can understand the hereditary information contained within the mRNA molecule and translate this information into amino acids. Exactly how does this translation occur? At one end of the tRNA molecule is a unique sequence of three ribonucleotide bases. This sequence, called an **anticodon,** recognizes its complementary codon in the mRNA molecule. Because each of sixty-one anticodons corresponds to a particular codon and because each of these codons corresponds to a particular amino acid (remember, the other three codons specify the stop signal), each tRNA anticodon must also correspond to a particular amino acid.

Elsewhere on the tRNA molecule is an enzyme which is able to bind the amino acid specified by the anticodon to the tRNA molecule in a reversible fashion. Each of the twenty binding enzymes specifies a single amino acid. For example, let's look at a tRNA molecule with an anticodon of CCC. This anticodon, complementary to an mRNA codon of GGG, specifies the amino acid glycine. Therefore, glycine is reversibly bound to this tRNA by the action of one of the specific enzymes. The tRNA molecule, along with its amino acid passenger, then travels to the ribosome where the amino acid becomes part of a growing polypeptide chain (Figure 2.10).

Translation: Getting from RNA to Polypeptide

We have talked about transcription—the process whereby all of the hereditary information contained in the genes of an organism is copied onto lengths of RNA. But the story does not end there. That hereditary information must then be converted into all of the many and varied polypeptide products that go on to define the final functions and structures of the organism. Some of those polypeptides function as enzymes that help certain chemical reactions to occur. For example, the enzyme called **lactase** helps you digest milk products so that you

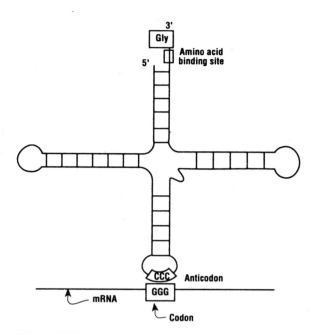

Figure 2.10
Characteristic two dimensional structure of a transfer RNA molecule. The amino acid binding site is located at the 3' end of the molecule. The anti-codon is located in the middle loop. In this illustration the anti-codon is represented by the base sequence CCC which is complementary to the codon GGG. GGG specifies the amino acid glycine which can be seen at the amino acid binding site.

will be able to eat that ice cream at lunch time. Other polypeptides become structural proteins such as those that form your hair, or your muscles. Still other proteins transport vital nutrients to the cells of your body and yet another set of proteins plays a role in the mechanisms of cell repair.

These various proteins differ from one another by the order of amino acids that make up each individual molecule. This order, known as the **primary structure** of the protein, is specified by information transferred from DNA to RNA during transcription. Translation—the process of protein synthesis—takes place in three main stages: **initiation, elongation,** and **termination** (Figure 2.11). Each of these steps involves specific enzymes, RNA molecules, ribosomes, and amino acids. Let's look at each step individually.

STEP ONE OF TRANSLATION: INITIATION The initiation of protein synthesis correctly places the mRNA on the ribosome. This occurs when the appropriate tRNA molecule binds to the AUG initiation codon on the mRNA molecule. Proper positioning of the mRNA is of vital importance in order to determine the **reading frame** of the message. The reading frame defines which groups of three bases

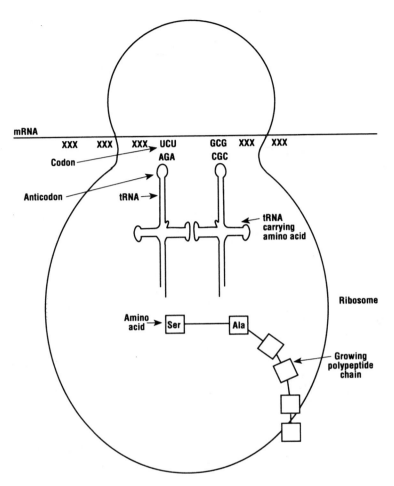

Figure 2.11
The translation of RNA into a polypeptide product takes place in three basic steps. **Initiation** involves the correct placement of the mRNA molecule on the ribosome. **Elongation** involves the attachment of amino acids—whose placement is dictated by the complementarity of the codon and anti-codon sequences—to one another to form a growing polypeptide chain. **Termination** involves the detachment of the completed polypeptide chain from the ribosome complex.

are to be read as codons. For example, the sequence AAUUCCUGG could be read in three different ways, each one of which would result in a different string of amino acids:

AAU UCC UGG

Asn Ser Trp

or

A AUU CCU GG

* Ile Pro **

or

AA UUC CUG G

** Phe Leu *

Because the very structure and function of an organism depends on its particular set of polypeptide molecules, such a situation obviously could not be successful! Instead, the presence of the start codon AUG, which encodes the amino acid methionine, signals the beginning of a genetic message. Only after methionine has been placed at the beginning of the message will translation progress. This insures that the reading frame will be correctly aligned and read. The correct alignment of the mRNA molecule and the ribosome is aided by the binding of a short sequence of ribonucleotide bases on the mRNA molecule to a complementary sequence located on the ribosome itself. The mRNA sequence, known as the **Shine-Dalgarno sequence** or the **ribosome binding site,** will be discussed in more detail in later chapters.

STEP TWO OF TRANSLATION: ELONGATION After the reading frame of the mRNA has been correctly aligned in the process of initiation, the second stage—elongation—begins. During elongation, additional amino acids are added to the growing polypeptide chain. Each new amino acid is brought to the mRNA-ribosome complex by a tRNA molecule whose anticodon is complementary to a codon on the message. The formation of a **peptide bond** joins the new amino acid to the previous amino acid. This type of bond gives the name "polypeptide" to the growing chain. Two things occur after formation of the peptide bond:

   1.   The tRNA, which carried the recently added amino acid, is released from the mRNA and the growing polypeptide chain. This tRNA will be recharged by the addition of another amino acid, and will continue to be active during the elongation of this or other polypeptides.

   2.   The position of the ribosome on the mRNA molecule shifts by the distance of one codon in a process known as **translocation,** thereby exposing a new codon to the various cellular components which play a role in translation.

Once a new codon is exposed the entire process begins again and continues until the entire genetic message has been translated into a polypeptide molecule, whose primary structure directly depends on the sequence of codons in the message. As elongation continues, the nascent polypeptide extends from the ribosome until the final step.

STEP THREE OF TRANSLATION: TERMINATION The end of polypeptide synthesis, or termination, occurs when translocation exposes a stop codon in the mRNA. This codon acts as the final punctuation in the message of the mRNA by telling the cellular machinery that the product is complete and to stop adding amino acids. At this point the completed polypeptide is released from the ribosome. In addition, the ribosomal subunits dissociate and separate from the mRNA. As translation ends, the mRNA molecule is either recycled for synthesis of additional polypeptide molecules or is broken down and degraded.

REVIEW AND SUMMARY

1. The phrase "one gene–one polypeptide" describes a series of experiments which demonstrated that each individual gene is responsible for directing the production of a single polypeptide.

2. The Central Dogma refers to the fact that hereditary information most often flows in one direction only—from DNA to RNA to polypeptide. The process in which information is transferred from DNA to RNA is called transcription, while the process of polypeptide synthesis is called translation.

3. Of the four types of RNA molecules, **mRNA** carries the hereditary message, **rRNA** is a structural part of the ribosome, **tRNA** "reads" the genetic code and carries amino acids to the polypeptide chain being synthesized on the ribosome, and **snRNA** plays a role in the conversion of immature mRNA molecules to mature mRNA molecules.

4. Complementarity plays a role in all stages of information transfer. Accurate transcription relies on the complementarity of deoxyribonucleotides and ribonucleotides. Accurate translation relies on the complementarity of the mRNA's codon and the tRNA's anticodon.

5. Transcription involves four steps: the **unwinding** of the DNA molecule to free the genetic template, the **pairing** of the template deoxyribonucleotide bases with complementary free ribonucleotide bases, the **joining** of adjacent ribonucleotides, and the **separation** of the new RNA molecule from the DNA template.

6. RNA is often processed or modified after transcription. This processing involves addition of a 3′ poly(A) tail and a 5′ cap structure. In addition, non-coding sequences called introns are spliced out of the RNA molecule, leaving behind only the coding exons.

7. The genetic code has an alphabet of four letters—A, U, C, and G. The words of the genetic code are each three letters long and are called codons. Although many amino acids are specified by more than one codon, no single codon specifies more than one amino acid. The beginning of a series of codons is defined by the reading frame which shows where each "word" of the genetic code starts.

8. Translation involves three steps: during **initiation** a tRNA molecule binds to the start codon to define the reading frame, during **elongation** the tRNA molecules "read" the message in the mRNA and transport amino acids to the appropriate positions, and **termination** occurs when a stop codon is reached. At termination, the new polypeptide is released from the ribosome which is then degraded into its subunits. The mRNA molecule is either recycled for further polypeptide synthesis or is broken down.

When something goes wrong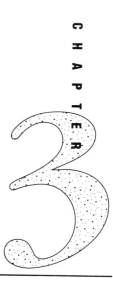

On August 6th, 1945, an atomic bomb was detonated directly over the Japanese city of Hiroshima (Figure 3.1). Conservative estimates suggest that a total of 70,000 people were killed immediately as a direct result of the blast. However, some people survived. Those survivors represented a heterogeneous slice of society—pregnant women, young children, adults of childbearing age, and older adults.

One of the survivors, a 26-year-old woman named Fumiko, became quite ill immediately after the bombing. She was nauseous and had diarrhea. Her skin became red and dry, and much of her hair fell out. To Fumiko's surprise, she survived despite these symptoms, and as time went on, her life returned to a sort of routine. Then, in 1952, Fumiko developed leukemia. She thought that it was an unfortunate coincidence that some of her friends also developed the disease and, as time went on, she heard of many other cases throughout the city. Fumiko died in December, 1952.

Studies of survivors of the Hiroshima bombing show that what happened to Fumiko was not an isolated occurrence; people exposed to the radiation from the bomb have developed various types of cancers at rates far higher than the general public. These rates continued to be elevated even into the 1980s. Scientists have used microscopic examination to look at chromosomes taken from the survivors. These examinations show evidence of unusually high levels of chromosomal breakage, resulting in chromosome fragments and other structural abnormalities. Although we may never know for sure what caused the cancers, it seems likely that exposure to radiation from the bomb resulted in changes to the body's normal genetic makeup, as well as structural abnormalities in the chromosomes themselves. What exactly are the genetic changes that may have occurred, and how did they arise?

42 When Something Goes Wrong

Figure 3.1
Hiroshima, Japan. August 6th, 1945. (a) An atomic bomb is detonated directly over the city. (b) Only the shells of a few buildings remain standing in the razed city.

OVERVIEW Cells must divide if growth is to occur. You started out as a single fertilized cell and, as a human adult, have grown into an organism made up of approximately ten trillion cells! Obviously this growth involved many, many cell divisions. Any time an individual cell divides, it must use the process of DNA replication to make a copy of the genetic material contained in its nucleus (Chapter One). During replication each strand of the DNA double helix serves as a template to direct the synthesis of a new complementary strand, thus yielding two, new, identical double-helix molecules. When the entire set of chromosomes in a cell—called the **genetic complement**—undergoes replication, it results in two full sets of chromosomes contained within the single parent cell. When this single parent cell divides into two new offspring cells, each of the new cells receives one of the two genetic complements. Each chromosome in one of the new cells has a base sequence exactly identical to the chromosomal base sequence in the other new cell. That is, the genetic complements of the two new cells are identical to one another.

Sometimes, however, a mistake is made involving the DNA, and the two genetic complements are no longer exactly equal in either content or structure. These mistakes, called **mutations,** are defined simply as changes in the base sequence of a DNA molecule. The processes of DNA replication and cell division transmit these changes along with the hereditary information contained in unchanged base sequences to the next generation of cells. This chapter examines the different types of mutations which can occur. In addition, it discusses the effects of mutations on the processes of transcription and translation, and answers the following questions:

1. What are the major types of mutations, and how do they affect the structure of the chromosome?

2. What are some of the causes of genetic mutations, and are all mutations harmful?

3. Can mutations be caused in a predictable way?

MUTATIONS: GENETIC MISTAKES

Copying Processes Are Not Always 100% Accurate

Think about the task of hand-copying all of the phone numbers from a single page of the telephone directory. Most of us would be unable to finish the task with complete accuracy. A pair of numbers might be reversed or left out entirely, or a number could be inverted or even scrambled beyond all recognition. Yet the process of replicating the DNA in a single human cell resembles copying not only a single page of the phone directory, but rather all of the residential phone numbers in the entire New York City phone book over and over, a total of 123 times! Yet the task of copying base sequences in DNA replication is an astonishingly accurate process.

The actions of the enzyme known as DNA polymerase in combination with a group of more than twenty DNA repair enzymes makes this accuracy possible. Together, the polymerase and repair enzymes act as a proofreading mechanism. Just as you would read a typed sentence and correct any of your mistakes, these enzymes follow along the newly replicated DNA strand and correct any mistakes—usually in the form of mismatched bases—that they detect. However, despite the action of the proofreading enzymes, mistakes, or mutations, do sometimes occur: estimates suggest that approximately three mistakes appear in every 3×10^9 deoxyribonucleotides which are replicated. In other words, only three mistakes would appear in the entire set of 123 copies of the phone book!

TWO BASIC TYPES OF MUTATIONS

Genetic mutations range from the involvement of a single base pair to the involvement of an entire chromosome and are broadly divided into two classes. One class, called **chromosomal mutations,** involves modifications of large sections of a chromosome or even the entire chromosome. The other type of mutation, called a **gene mutation,** involves only a single gene or a small fragment of a chromosome.

Chromosomal Mutations: Changes in Number

Most living organisms contain genetic information, stored in the base sequence of the deoxyribonucleotides which make up the chromosomal DNA. Each organism has a certain number of chromosomes in which to store all of the information that it needs, and, although there are exceptions, that number remains constant among all organisms of any particular species. In other words, all normal humans have forty-six chromosomes which contain all of the genetic material needed to act as the blueprint for a single human being. All normal mice, on the other hand, have only forty chromosomes, while houseflies have twelve and mosquitoes have six chromosomes. Interestingly, the seeming complexity of an organism appears to be unrelated to the number of chromosomes possessed by the organism—some flowers have more than one hundred chromosomes!

One type of chromosomal mutation, called an **aneuploid mutation,** involves a change in the number of chromosomes characteristic of a given species. The condition referred to as **Down's syndrome** is an example of an aneuploid mutation. A person with Down's syndrome has forty-seven chromosomes rather than the normal complement of forty-six (Figure 3.2). The extra chromosome is called chromosome 21 because of its size in relation to the other human chromosomes. Down's syndrome, sometimes referred to as **Trisomy 21,** a term meaning three copies (*tri*somy) of chromosome 21, occurs in approximately one out of six hundred live human births. A Down's syndrome individual generally has an IQ of less than 70, as well as numerous phenotypic abnormalities, including the characteristic epicanthal fold of the eyelid.

In contrast to Down's syndrome in which there are too many chromosomes, aneuploid mutations can also involve a loss of chromosomes. A person with **Turner's syndrome,** an example of this type of aneuploid mutation, has only forty-five chromosomes due to the absence of one X chromosome. Such a person is a phenotypic female with a normal IQ. Phenotypic abnormalities include immature secondary sex characteristics, a webbed neck, and short stature.

While numerous examples of other human aneuploid mutations exist, those surviving to adulthood generally involve only chromosome 21, or the sex chromosomes, X and Y. Other aneuploid mutations, with rare exceptions, are lethal, causing either fetal death or a severely shortened lifespan.

Figure 3.2
A person with Down's syndrome has three copies of chromosome 21, instead of the normal two copies. This aneuploid mutation results in a variety of effects including a characteristic facial appearance and mental retardation.

Chromosomal Mutations: Changes in Structure

Other mutations can occur during **meiosis,** a series of two cell divisions in which a single cell containing pairs of chromosomes divides to produce four cells, each of which contains only a single member of each chromosome pair (Figure 3.3). During meiosis, chromosomes similar to one another in size and structure, called **homologs** or **homologous chromosomes,** pair up so that they are lying side by side. During this process of pairing up, called **synapsis,** bridges form between the two homologs and pieces of chromosomes are sometimes exchanged. Since the exchanged pieces are usually equal in size, each of the two homologs ends up with the same number of genes with which it started. This process is called **crossing over.** Crossing over is technically not a mutation, since genetic information remains unchanged by the switch. The process does, however, have the potential to create new combinations of phenotypes (Figure 3.4). Sometimes, however, the pieces that are exchanged during crossing over are unequal. When this occurs, one chromosome may end up with two copies of one or more genes, and one chromosome may end up missing certain genes entirely.

Sometimes exchanges of genetic material occur between two non-homologous chromosomes as a result of breaks somewhere along the length of one or more chromosomes. Although the breaks are often rendered harmless by quick and correct repair, inadequate or incorrect repair may result in a change in the overall structure of one or more individual chromosomes. These types of chromosomal mutations, divided into categories by the kind of structural damage which has occurred, follow:

1. **Translocations** involve the movement of a chromosomal piece or pieces. **Simple** translocations involve a break in one chromosome to create a detached segment which is then joined to a new chromosome. This process yields one chromosome which is shorter than normal and one which is longer. **Reciprocal** translocations involve breaks in two chromosomes. The

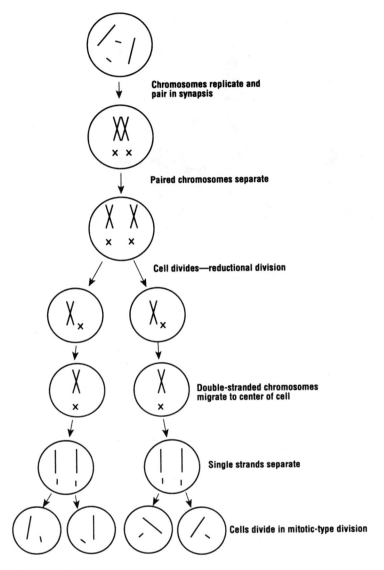

Figure 3.3
Overview of meiosis—a process of cell division which results in the production of four daughter cells from a single parent cell. Each daughter cell contains only half the number of chromosomes that was contained in the original parent cell.

segments then switch so that segment A joins to chromosome B while segment B joins to chromosome A. Reciprocal translocations differ from crossing-over events in that the former involve non-homologous chromosomes, while the latter involve homologous chromosomes. Genetic material is neither gained nor lost in either crossing over events or in translocations (Figure 3.5a). Although some types of translocations have minimal effect, other types have more serious results. One such translocation is referred to as the **Philadelphia translocation,** named after the city in

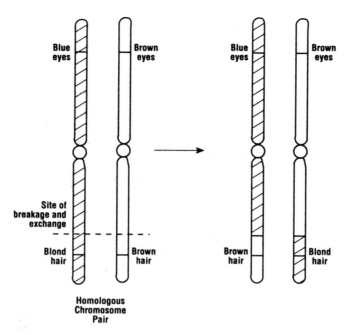

Figure 3.4
The exchange of genetic material between homologous chromosomes creates new phenotype combinations. Where originally blue eyes were paired with blond hair, and brown eyes with brown hair, crossing over results in the re-pairing of these phenotypes. Blue eyes are now paired with brown hair while brown eyes are paired with blond hair. This example is given for illustrative purposes only: in reality the genes which encode hair and eye color are not located on the same chromosome.

which it was first described. The **Philadelphia chromosome,** characterized by the translocation of a piece of chromosome 22 to another chromosomal location, is often associated with the development of chronic myelogenous leukemia and is found in many patients with this disease.

2. **Deletions** involve the breakage of a chromosome to produce a chromosomal fragment. If that fragment does not either reattach to the original chromosome or translocate to a new chromosome, it will be lost, or deleted, from the cell. If the deletion is large enough, it shows up microscopically by the presence of a chromosome which is shorter than normal. Copies of any genes located on the deleted piece are lost to the organism (Figure 3.5b). Such a deletion of a piece of chromosome 5 causes a set of abnormalities known as **Cri-du-chat syndrome.** Babies with this condition have a peculiar cat-like cry (which gives the syndrome its name), as well as low IQ, microcephaly and other distinctive phenotypic characteristics.

3. **Duplications** involve the copying of a chromosomal segment and result in production of a longer-than-normal chromosome. This chromosome has multiple copies of genes that would normally appear only once on a single chromosome (Figure 3.5b). There are no known cases of chromosomal duplications which have resulted in live human births.

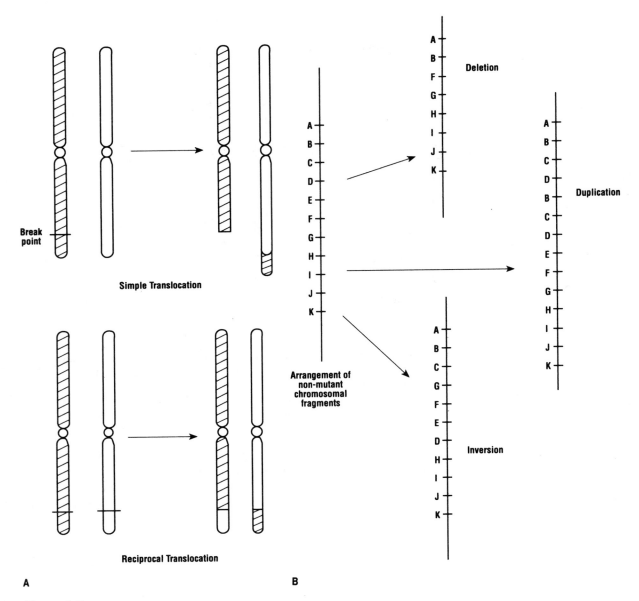

Figure 3.5
Alterations in chromosomal base sequence. Mutations can affect base sequences in four basic ways including (a) the trading or translocation of a fragment between two non-homologous chromosomes and (b) the duplication, deletion, or inversion of a fragment on a single chromosome. The letters in this Figure represent chromosomal fragments, each composed of a sequence of base pairs.

4. **Inversions** involve the reversal of a length of chromosome within the chromosomal structure (Figure 3.5b). Because inversions do not change the amount of DNA in a chromosome, they often have no phenotypic effect. In some cases however, the presence of an inverted segment interferes with the process of synapsis in meiosis. This interference may result in meiotic abnormalities which in turn may cause the formation of abnormal germ cells.

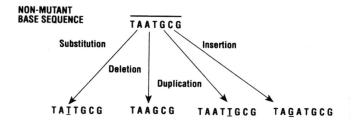

Figure 3.6
Point mutations involve changes in single base pairs within a nucleotide base sequence.

The effects of all of these changes in chromosomal structure can range from minor to catastrophic and lethal, depending on the extent of the structural changes. In general, changes affecting larger chromosomal fragments yield more damaging results than those that affect more limited areas.

Gene Mutations: Changes in Individual Genes

We have described mutations which change the structure of entire chromosomes. Many of these mutations do not, however, change the structure of individual genes. Although the number of copies of a gene may change, the actual genes themselves may be unaffected. There are, however, gene mutations which change the base pair sequence of individual genes. When these mutations involve only one base pair, they are called **point mutations.** Point mutations may involve the **substitution** of one base for another, the **deletion** of a base or bases, the **duplication** of one or more bases so that two identical bases exist where one base should be, or the **insertion** of one or more new unrelated bases into the base sequence (Figure 3.6).

MUTATIONS: NEW GENETIC MESSAGES AND THEIR CAUSES

Mutations Can Alter Polypeptides by Altering Codons

Genetic mutations can be categorized by their effects on individual codons. If a mutation alters a base within a codon so that the modified codon now specifies a different amino acid, it is called a **missense mutation.** If, however, a mutation changes a codon without changing the identity of the encoded amino acid, it is called a **same sense mutation.** For example, the codon UUA specifies the amino acid leucine. If this codon mutates to read UUG, it still encodes the amino acid leucine. This is a same sense mutation. A mutation that causes a codon to become a stop codon is called a **nonsense mutation.** An example of this would be the mutation of UUA, which encodes the amino acid leucine, to UAA which is a stop codon (Figure 3.7).

A mutation which involves the addition or deletion of one or two bases causes a change in reading frame (Chapter Two) and is called a **frameshift mutation.** Such a mutation changes the reading of every codon which occurs in the 3' direction, or **downstream,** of the mutation (Figure 3.8). Of course such a mutation greatly affects the polypeptide product of any specific gene.

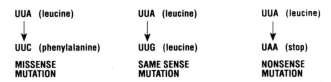

Figure 3.7
Point mutations can have a variety of effects on single codons. This diagram shows three possible results from the substitution of a single base within a codon. A missense mutation results in production of a codon which specifies a different amino acid than was specified by the unmutated codon. A same sense mutation produces a codon which specifies the same amino acid as was specified by the unmutated codon. A nonsense mutation results in the conversion of a codon which specifies an amino acid into a stop codon.

	DNA base sequence	mRNA base sequence	Amino acid sequence
Non-mutant	TTA AGC CCG GTA TGG	AAU UCG GGC CAU ACC	Asp Ser Gly His Thr
Mutant (point deletion)	TTA GCC CGG TAT GG * * *	AAU CGG GCC AUA CC * * *	Asp Arg Ala Ile ** * * * *

Figure 3.8
Frameshift mutations involve the deletion or insertion of one or two bases into the coding sequence of a gene. As a result, the codon reading frame is changed in the complementary mRNA molecule, thus altering the final polypeptide product. In this illustration ★ represents a codon and/or amino acid which differs from the original sequences as a result of mutation.

Mutations Can Cause New Phenotypes

We have seen that the arrangement of bases which make up a gene is sometimes altered by various types of mutations so that the new arrangement of bases is different from the old. From our discussion of the processes of transcription and translation, we know that a change in the base sequence of a gene results in a change in the base sequence of a molecule of mRNA. In turn, the change in the mRNA sequence results in a change in the sequence of amino acids which make up the final polypeptide product. This cascade of changes—from DNA to RNA to polypeptide—results from the complementarity of the processes which govern the flow of hereditary information.

These alterations in the polypeptide product can lead to variations in phenotype, ranging from undetectable to cosmetic (as in a streak of unpigmented hair on the head of a brunette) to life threatening to lethal. In some cases, however, such alterations may be beneficial to the organism. An example of a potentially beneficial alteration is the point mutation involved in sickle-cell anemia. We know that this disease results from a single-point mutation in the hemoglobin gene. In order to have full-blown sickle-cell anemia, a person must have inherited this point mutation from both parents.

Figure 3.9
Mutagens cause genetic mutations. Cigarette smoke, a potent mutagen, contains more than 6500 different mutagenic compounds.

However, if a person inherits this point mutation in the hemoglobin gene from only one parent and the other hemoglobin gene is normal, a potentially beneficial effect can be derived from the sickle-cell anemia point mutation. That effect is resistance to the disease malaria!

Mutations Affect Both Germ Cells and Somatic Cells

Genetic mutations can affect either the **germ cells**—those giving rise to reproductive cells such as sperm and egg—or the **somatic cells,** which make up the non-reproductive tissues of an organism, such as cells of the liver or cornea. Mutations which affect the germ cells can cause **heritable** changes, passed along to future generations as part of the genetic complement. In contrast, somatic mutations affect only future generations of a cell within a single organism. It should be pointed out, however, that although somatic mutations are not passed on to successive generations, a tendency to develop somatic mutations may be inherited.

Mutations Arise From a Variety of Causes

Some mutations occur spontaneously as a result of errors made during the process of DNA replication. Mutations can also be caused by exposure to certain physical and chemical agents known as **mutagens** (Figure 3.9). These chemicals include such ubiquitous agents as cigarette smoke, which contains more than 6,500 different mutagenic compounds, ultraviolet radiation from the sun, chemicals in cleaning solvents and paints, and various common food additives, such as nitrites in smoked meats.

All mutations, regardless of their sources, are subject to the proofreading mechanism of the DNA repair enzymes and are often detected and corrected before they cause any noticeable effects on the phenotype of the organism. However, the presence of inherited genetic diseases such as sickle-cell anemia and the number of cancers which result from genetic mutation every year, make it apparent that not all mutations are corrected by the DNA repair enzymes.

Mutations Are Not Predictable

As we have seen, mutagens act to mutate DNA, and some of those mutations lead to altered phenotypes. Sometimes these mutations cause phenotypes that are beneficial to the organism, while other mutations cause harmful phenotypes or some variations of phenotypes that are insignificant. Until recently no one had any control over the process of mutation-induced phenotypic variation; it was random. Scientists quickly realized the benefits of controlling the process of mutation to create mutations that have specific phenotypic effects. Before such a feat becomes possible, however, we must be able to characterize and manipulate chromosomes and the genes which are found on them.

REVIEW AND SUMMARY

1. Mutations are changes in hereditary information and can affect the DNA of both germ cells and somatic cells.

2. Point or gene mutations involve changes in the base sequence of individual genes. These mutations may involve the **substitution, deletion, duplication,** or **insertion** of a base or bases, or the **inversion** of a short sequence of bases.

3. Chromosomal mutations are changes in the overall structure of a chromosome involving the **translocation, deletion, duplication,** or **inversion** of a chromosomal fragment or fragments.

4. All mutations have the potential of altering one or more codons. A **missense** mutation arises when the DNA is altered so that it specifies a "wrong" codon which, in turn, specifies a "wrong" amino acid. A **same sense** mutation results when the altered codon continues to specify the "correct" amino acid. A **nonsense** mutation arises when a codon which specifies an amino acid becomes a "stop" codon. A **reading frame shift,** a change in the position of breaks between codons, results from the insertion or deletion of one or two bases in a gene.

5. An altered codon may encode a "wrong" amino acid which, in turn, results in the translation of a "wrong" polypeptide product. Therefore, mutations have the potential to alter genetic products and resultant phenotypes.

6. The chromosomes of both germ cells and somatic cells can be affected by mutation. Mutations involving germ cells can be passed on to the next generation, while mutations of somatic cells only affect future generations of the individual cell. Offspring will not be affected by somatic cell mutations, although they may inherit the tendency to develop somatic cell mutations.

7. Mutations can be caused by a variety of mutagenic agents, including elements contained in cigarette smoke, food additives, and environmental contaminants. In addition, some mutations arise simply from complications in the process of cell division.

8. Most mutations are corrected by the proofreading mechanisms of the cell before they cause any phenotypic abnormalities. The proofreading mechanism consists of a series of enzymes which attempt to detect and repair mismatched bases of the deoxyribonucleotides in the DNA molecule.

Isolating and manipulating DNA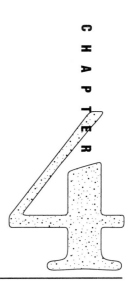

Pauline D., a 24-year-old diabetic female living in the Seattle area, was diagnosed with diabetes in 1988, and she has been giving herself four shots of insulin—a pancreatic hormone that regulates sugar metabolism—each and every day since her diagnosis (Figure 4.1). She needs these insulin injections because her body does not produce any insulin of its own. The lives of Pauline and more than one million other diabetics in the United States depend on daily injections of insulin. Like many diabetics today, Pauline injects herself with a rather unusual insulin. She uses **human** insulin that is isolated from **bacterial** cells.

Until 1982 nearly all of the insulin used to treat diabetics was isolated from the pancreases of cows or pigs that had been slaughtered for food. This bovine and porcine insulin, while similar to human insulin, does have some small structural differences. Most diabetics found the animal insulin very effective in controlling their blood sugar levels. However, some people developed allergic reactions to this insulin from other species. They needed a source of *human* insulin to control their disease.

What doctors and scientists needed to solve this problem was the stuff of which science fiction is made: they needed a source of human insulin that wasn't human!

OVERVIEW In the usual scheme of things, human cells contain human DNA and produce human proteins, while bacterial cells contain bacterial DNA and produce bacterial proteins. It seems apparent that for any cell to synthesize a particular protein that cell must contain the gene that encodes that protein. Therefore, bacterial cells containing no human DNA produce no human proteins. But what if we want to make a bacterial cell which *does* produce human proteins? In order to accomplish this remarkable goal, we must find a way to put human DNA into that bacterial cell. The first step toward this goal is the isolation and manipulation of actual cellular DNA, because only then can DNA be moved from one cell to another at will.

Isolating and Manipulating DNA

Figure 4.1
This young woman like Pauline measures insulin in preparation for giving herself the first of four daily injections which she needs to maintain good health.

This chapter addresses these issues and answers the following questions:

1. How is DNA contained within an individual cell?
2. How do scientists remove DNA from the interior of a cell?
3. What special tools do scientists use to cut and paste the fragile, microscopic DNA molecules?
4. What is a recombinant DNA molecule?

CELLS AND THEIR HEREDITARY MATERIAL

All Organisms Are Composed of Cells

The fundamental unit of all living organisms is the **cell.** Cells range in size from the smallest bacterium, which may be only about 0.2 microns (1 micron = 1/1,000,000 of a meter) in length, to the egg of an ostrich, which may be 7 cm, or almost three inches, across. Most of the cells of your body are intermediate in size and measure about 20 microns in length.

All cells are divided into two basic categories: **eukaryotic** and **prokaryotic** cells. The presence or absence of a **nucleus** (plural: **nuclei**)—a membrane-bound area of the cell that contains the cell's hereditary material—distinguishes these two cell types. Eukaryotic cells have distinct nuclei and include the cells of humans, animals, and most varieties of plants (Figure 4.2). Prokaryotic cells do not contain nuclei. Instead, the hereditary material of these cells is found in an area, called the **nucleoid area**, which is not membrane-bound (Figure 4.3). Prokaryotic cells compose bacteria and **cyanobacteria**, sometimes called blue-green algae.

Cells of All Organisms Contain the Same Genetic Elements

We have already seen in Chapter Two that the detailed genetic instructions contained in the nucleic acids of the cells determine the characteristics of all living organisms. We also learned that only five different subunits, called nucleotides, compose the DNA and RNA molecules of all organisms and that these five nucleotides closely resemble one another in their chemical structures (Figures 1.8). Each individual gene is characterized by the arrangement of its nucleotides in a particular, unique, sequence and different genes have different sequences of nucleotides. Relatively small differences in the nucleotide sequence of genetic material often results in the production of organisms as widely different as people, pears, paramecia, and petunias. For example, there is only a few percent difference in the nucleotide sequences of the DNA of apes and of humans, yet we know that this small variation leads to remarkable differences in structure and ability between these two species (Figure 4.4).

DNA IS ARRANGED IN CHROMOSOMES

All organisms, whether prokaryotic or eukaryotic, contain their hereditary information in the form of DNA stored inside the cell. DNA is divided into functional subunits called genes (Chapter 2), and cells have a great many genes. A typical bacterial cell contains approximately 3,000 genes, while a human cell may have as many as 100,000 genes which are composed of roughly 3×10^9 deoxyribonucleotide pairs. If the DNA in an average human cell were stretched out, it would be about three meters—or more than nine feet—in length! How does the relatively small interior of the cell store all of this DNA?

Compact structures called **chromosomes** store DNA inside the cell. Each chromosome is thought to consist of a single strand of double helix DNA wrapped around special proteins called **histone** proteins. These proteins act as spools to hold the DNA, much as spools hold the thread that you might use to mend a torn shirt. Each individual histone protein complex holds a length of DNA consisting of 140 deoxyribonucleotide pairs. The histone/DNA complex is called a **nucleosome.** A single molecule of DNA is arranged into multiple nucleosomes which in turn are coiled and looped on themselves to form a substance called **chromatin.** The compact chromatin is then coiled and folded to form an even more compact structure that we know as a chromosome (Figure 4.5). All of this precise coiling and folding enables the equivalent of three meters of DNA to fit inside a single microscopic cell!

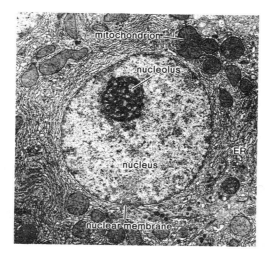

Figure 4.2
The eukaryotic cell, found in animals—including humans—and most varieties of plants, contains a membrane-bound nucleus along with many other cellular organelles.

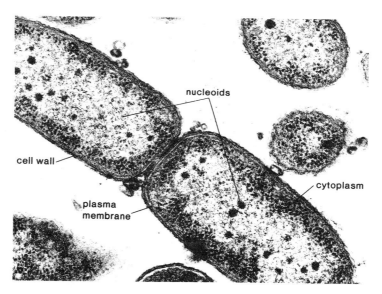

Figure 4.3
The prokaryotic cell, unique to bacteria and cyanobacteria, is characterized by the absence of a membrane-bound nucleus.

Chromosomes Can Be Either Circular or Linear

Eukaryotic and prokaryotic cells differ somewhat in the structure and placement of their chromosomes. Eukaryotic cells generally have multiple chromosomes arranged as long, thin, linear structures found inside the nuclear membrane. Conversely, prokaryotic cells have a single circular chromosome which is not contained within a nucleus. In addition, prokaryotic cells frequently have one or more **plasmids**—circular pieces of DNA which are separate from the chromosome. Plasmids often contain genes which give a cell the ability to survive

A

B

Figure 4.4a & b
The deoxyribonucleotide sequence of the DNA molecules of humans and apes differs by only a few percent. This small difference is, however, sufficient to produce species widely divergent in both physical appearance and abilities, as well as in mental capacities.

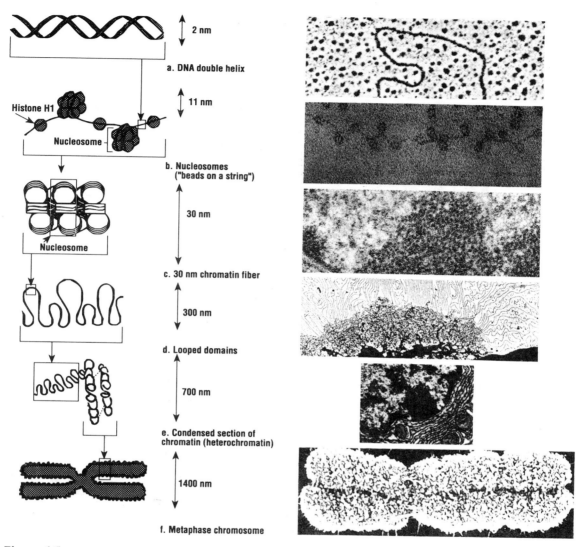

Figure 4.5
DNA is packaged in a series of steps to form chromosomes. Each double-stranded DNA molecule is wrapped around histone proteins to form a nucleosome. The nucleosomes, arrayed along the length of the DNA molecule, are coiled on themselves to form a chromatin fiber which, in turn, loops and coils on itself to form a compact chromosome.

in the presence of materials normally toxic to that cell. For example, some plasmids contain a gene which enables bacterial cells containing those plasmids to survive in a fluid supplemented with penicillin, despite the fact that bacterial cells are generally unable to survive in the presence of this drug. Such plasmids are said to confer **drug-resistance** or **antibiotic-resistance** on cells which contain them.

Figure 4.6
Strands of DNA precipitate out of an aqueous solution when they come in contact with ethanol. Ethanol layered on top of an aqueous DNA solution forms an interface at which the DNA will precipitate. Precipitated DNA—which appears as white threads—can be collected by "spooling" on a glass rod.

DNA ISOLATION: "GENOMIC" OR CHROMOSOMAL DNA

The chromosomes of eukaryotic cells, sometimes referred to collectively as **genomic DNA,** can be isolated using relatively simple chemical techniques. These techniques rely on the fact that the major component of all cell membranes is a fatty molecule called a **lipid.** Lipids decompose or dissolve when they are treated with detergents. This important property is demonstrated every time you clean a greasy dish by washing it with a mixture of dishwashing detergent and water. The detergent dissolves or **solubilizes** the grease so that it can be washed down the drain. Cell membranes treated with detergent solubilize in much the same way.

A solubilized cell membrane develops holes which allow the contents of the cell to leak out. This leaked material, called the **lysate,** consists of proteins, RNA, and cell debris, along with the DNA which is to be isolated. In order to separate the DNA from the contaminating cellular materials and collect it in as pure a form as possible, the lysate is treated with a variety of enzymes to digest the contaminants. These enzymes typically include a **protease** to digest protein molecules and an **RNase** to digest RNA molecules. In addition, the lysate is heated to 60°C (140°F) in order to inactivate any DNA-degrading enzymes which may be present in the lysate. This heating process, known as **heat inactivation,** is effective because most protein molecules, including enzymes, lose the structure necessary for their proper functioning—that is, they **denature**—at 60°C. DNA itself does not denature until a temperature of 80°C (176°F) has been reached. This means that the appropriate application of heat can destroy contaminating enzymes while leaving the desired DNA molecules intact.

Collection of the purified DNA from the enzyme and heat-treated lysate relies on the fact that DNA molecules are soluble in water, but are not soluble in a type of alcohol called **ethanol.** When ethanol is layered on top of a water-based—or **aqueous**—solution containing DNA, the different densities of the two liquids causes an obvious border, called an **interface,** to form between the water and the ethanol (Figure 4.6). Any DNA present at this interface will **precipitate,** or come out of, the aqueous solution. Precipitated DNA appears as long, thin, white threads gathered into a cloudy-looking mass and can be collected from the interface by inserting a glass rod into the DNA precipitate. As the rod is twisted, the precipitated DNA threads will wind or spool around the rod. This spooled DNA can be dried, dissolved in solution, and used in a variety of procedures.

DNA ISOLATION: BACTERIAL PLASMIDS

You will remember that a plasmid is a circular piece of DNA, separate from the chromosomal DNA, often found in bacterial cells. The isolation of bacterial plasmid DNA poses a unique problem since it involves the separation of two different types of DNA: the plasmid DNA must be separated from the chromosomal DNA before it can be purified. In contrast, the isolation of eukaryotic genomic DNA involves no such separation because eukaryotic cells do not contain plasmids along with the chromosomal genomic DNA.

As with eukaryotic cells, treatment of the prokaryotic bacterial cell with detergent causes the cell membrane to solubilize, thus releasing both plasmid DNA and chromosomal DNA along with various contaminating molecules. The chromosomal DNA, along with some protein molecules, is precipitated out of this lysate by adding potassium to the lysate, often in the form of a potassium acetate/acetic acid solution. This chromosomal DNA/protein/potassium precipitate can be removed from the lysate by **centrifugation** (spinning at high speeds). Centrifugation causes the precipitated material to move to the bottom of the centrifuge tube just as swinging a bucket vigorously in circles at arm's length causes a ball to remain firmly in place, pressed against the bottom of the bucket (Figure 4.7). The precipitated chromosomal DNA/protein/potassium complex can then be separated from the remaining lysate leaving a **cleared lysate.** The cleared lysate contains the plasmid DNA along with contaminating RNA and protein and perhaps a small amount of chromosomal DNA debris. The RNA can be removed by digesting it with the enzyme RNase, and the proteins, including the RNase enzyme, can be removed by mixing phenol with the cleared lysate. Since phenol and the cleared lysate react with each other in much the same was as oil and water, they will not remain mixed together. Instead, they separate into layers, with the phenol layer containing the contaminating proteins and the water-based lysate layer containing the plasmid DNA. Once the phenol layer is removed, the plasmid DNA can be precipitated out of the water-based layer with alcohol in much the same way as genomic DNA is precipitated.

Once DNA molecules are collected, the genetic engineer must be able to manipulate this DNA. Only after an isolated DNA molecule, be it chromosomal or plasmid in origin, can be "cut and pasted" according to the specific demands of the scientist, can it be used to alter the genetic characteristics of various cells. Examples of this kind of work include the cutting and pasting of human and bacterial DNA that caused bacterial cells to make human insulin. The following sections discuss some of the most common procedures used for these types of DNA manipulations.

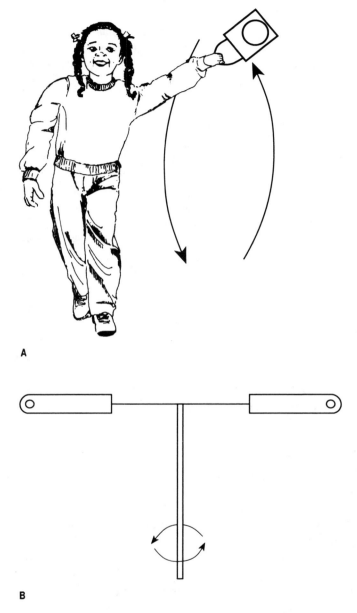

Figure 4.7
A **centrifuge** is an instrument capable of spinning materials in test tubes at speeds of up to 100,000 revolutions per minute. (a) A ball in a spinning bucket is forced to the bottom by centrifugal force, and thus doesn't fall from the open mouth of the bucket. (b) The centrifugal forces which result from the spinning process also cause particulate matter to be forced toward the bottom of the test tubes.

RESTRICTION NUCLEASES

DNA Molecules Are Cut by Bacterial Enzymes

Toward the end of the 1960s, scientists Stewart Linn and Werner Arber, carrying out research in Geneva, hoped to isolate an enzyme from bacterial cells which would have the ability to modify DNA structure. Using a strain of bacteria called ***Escherichia Coli b*** (***E. coli***) they isolated two enzymes with very interesting properties. One of those enzymes functioned to add certain chemical groups, called **methyl groups,** to DNA, while the other enzyme seemed to cut or **cleave** the sugar-phosphate backbone of unmethylated DNA or DNA which was lacking in methyl groups. This DNA cleavage meant that a length of DNA could be broken up into two or more smaller fragments, just like using a pair of scissors to cut a long rope into smaller pieces.

Scientists immediately recognized the value of an enzyme able to break long pieces of DNA into shorter fragments, particularly if that enzyme could be made to cut DNA at specific, predetermined places. The scientists wanted an enzyme that would "read" the sequence of base pairs in a DNA molecule and then cut that molecule at specific places. Those places would be determined by the actual base sequence of the DNA nucleotides. It was predicted that an enzyme which cut DNA molecules at sites which were determined by the base sequence of the DNA would be an invaluable tool in the manipulation of DNA molecules.

Unfortunately, those first enzymes which were isolated from *E. coli* cut the DNA molecule at random sites which appeared to be completely unrelated to the base sequence of the molecule.

Enzymes Are Found Which Cut DNA at Specific Sites

It wasn't until the following year, 1970, that the first sequence-dependent cleavage enzyme was discovered. Hamilton Smith, working at Johns Hopkins University, isolated this important enzyme from a strain of bacteria called **Haemophilus influenzae.** This enzyme, which was given the name **HindII,** always cut a DNA molecule at a certain point within a specific sequence of six base pairs. This base sequence is:

$$5'G\ T\ (T\ or\ C)\ (A\ or\ G)\ A\ C\ 3'$$
$$3'C\ A\ (A\ or\ G)\ (T\ or\ C)\ T\ G\ 5'$$

*Hin*dII always cuts directly in the center of this sequence. Wherever this particular sequence of six base pairs occurs in a length of DNA, *Hin*dII will cleave the DNA backbones between the middle two bases. Furthermore, *Hin*dII will *only* cleave DNA at this particular sequence of base pairs. If the sequence is not present, the *Hin*dII enzyme remains inactive, and the DNA molecule will remain intact. Because *Hin*dII only functions after recognizing this specific base sequence, the sequence is called a **recognition sequence.**

*Hin*dII belongs to a class of enzymes known as **restriction nucleases.** *Hin*dII is more specifically known as an **endo**nuclease because it cuts the DNA backbone at an internal site on the molecule. Other restriction enzymes, known as **exo**nucleases, require a "free end" on the DNA molecule. In other words, the molecule must be linear, thus having two ends, rather than circular. Circular molecules have no nucleotides which are not bound on both sides by other nucleotides and, therefore, have no free ends. Exonuclease enzymes work to degrade the DNA, one nucleotide at a time, by cutting the **terminal** or end nucleotide away from the rest of the sequence (Figure 4.8).

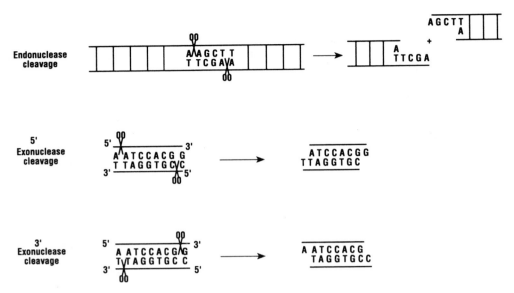

Figure 4.8
Restriction nucleases cleave DNA molecules in specific fashions as determined by the category of nuclease. Restriction endonucleases cleave within the molecule after recognizing specific base sequences known as **recognition sequences.** The 3' restriction exonucleases cleave and remove terminal bases at the 3' end of the DNA strand, while 5' exonucleases remove terminal bases at the 5' end.

All Restriction Nucleases Have Specific Recognition Sequences

All restriction endonucleases begin the cleavage process by recognizing specific recognition sequences within the DNA molecule. These sequences are generally four to six base pairs in length and are often symmetrical or palindromic—that is, the bases are the same on both DNA strands when each strand is read in the same direction (e.g., 3' to 5'. Remember that DNA strands are anti-parallel). (Chapter 1). DNA palindromes get their name from linguistic palindromes in which a sentence or phrase reads the same both backwards and forwards. For example, "MADAM I'M ADAM" or the more complex "ABLE WAS I ERE I SAW ELBA" are both linguistic palindromes.

Each individual enzyme has only one specific recognition sequence. However, some enzymes share recognition sequences with one another. In other words, two or more different enzymes may be specific for the same recognition sequence. Such pairs of enzymes are called **isoschizomers.**

When a restriction endonuclease encounters its particular recognition sequence, it cleaves the covalent bonds which link adjacent deoxyribonucleotides in the DNA molecule. This results in a break in the DNA backbone. For example, **simian virus 40,** also known as **SV40,** normally infects monkey cells. The virus has a double-stranded circular chromosome composed of 5,226 deoxyribonucleotide pairs. The base sequence "GGATCC" (Note: All recognition sequences in this text will be written as single DNA strands in the 5' to 3' direction. Recognition sequences are double stranded. The second strand of the recognition sequence, which is always both complementary and anti-parallel to the first, is

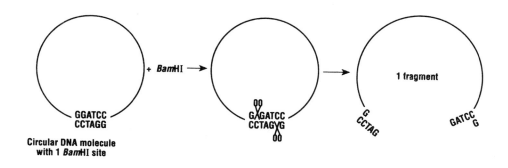

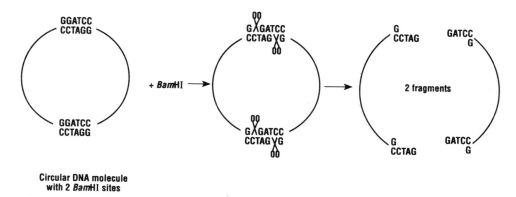

Figure 4.9
The number of fragments created when a DNA molecule is treated with a restriction enzyme is directly dependent on the number of specific recognition sequences present in that molecule.

not included in this text.) occurs a single time along the length of the SV40 chromosome. This sequence of bases is the recognition sequence for the restriction enzyme known as **Bam**HI. Therefore, when the circular SV40 chromosome is treated with the endonuclease BamHI, a single break will be made in the chromosome. This break converts the circular DNA molecule into a linear molecule. The more endonuclease recognition sequences there are in a DNA molecule, the more DNA fragments will be produced by enzymatic cleavage (Figure 4.9).

More than 900 restriction enzymes—some sequence specific and some not—have been isolated from approximately 230 different bacterial strains since the discovery of HindII in 1970. Since that time enzymes that recognize more than 130 different nucleotide recognition sequences have been identified.

Restriction enzymes, in general, have names that reflect the type of prokaryotic cell from which they are isolated. For instance, the endonuclease EcoRI is isolated from *Escherichia coli* RY13 bacterial cells, and the endonuclease HindII is isolated from *Haemophilus influenzae* Rd bacterial cells.

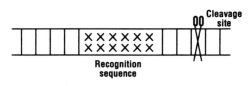
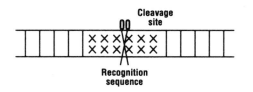

Figure 4.10
Type I restriction endonucleases bind to specific recognition sites but then cleave the DNA molecule at a random site generally located outside of the recognition site. In contrast, type II restriction endonucleases both bind and cleave within a specific recognition site.

Restriction Enzymes Protect Bacterial Cells in Nature

In nature, restriction endonucleases function to degrade any non-bacterial or "foreign" DNA which might have entered a bacterial cell. This type of event might occur when a virus infects a bacterial cell. A virus able to infect a bacterial cell is called a **bacteriophage,** or more simply, a **phage.** A phage infects a bacterial cell by injecting its DNA directly into the target or host cell so that the host cell contains both bacterial and phage DNA. In order to fight the infection, bacterial restriction enzymes work to cleave and degrade the invading phage DNA. The DNA of the bacterial cell is protected from cleavage by the restriction enzymes because this DNA contains chemical groups called **methyl groups,** which are attached to the adenine and/or cytosine bases along the length of the DNA molecule. The majority of restriction enzymes which have been isolated to date cannot cleave methylated DNA molecules. Invading bacteriophage DNA is not methylated at the recognition sequences specified by the host endonucleases and can therefore be degraded by the restriction enzymes.

TWO TYPES OF ENDONUCLEASES

Restriction endonucleases can be divided into **type I** and **type II** enzymes. Type I enzymes bind to the recognition sequence, but cut the DNA at essentially random sites along the length of the DNA molecule. Type II enzymes, on the other hand, cleave the DNA molecule at a specific site contained within the recognition sequence (Figure 4.10). It is the type II enzymes that are especially useful in genetic engineering. In fact, restriction nucleases have been exploited by scientists who use them as "biological scissors," and have proved essential in the development of genetic engineering techniques.

Restriction Enzymes Act as Molecular Scissors

Restriction enzymes are able to scan a length of DNA and read the DNA bases much as you are reading the words in this sentence. When the enzyme encounters its specific recognition sequence, it makes one cut in each of the two sugar-phosphate backbones of the DNA double helix. The positions of those two cuts,

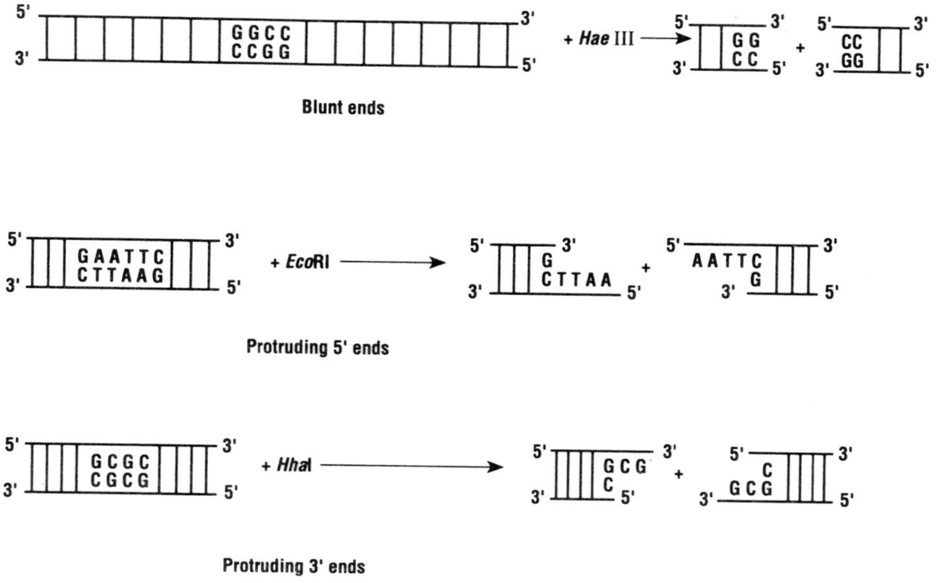

Figure 4.11

Cleavage by type II restriction endonucleases results in the production of one of three types of termini—blunt ends, protruding 5' ends, and protruding 3' ends—on the resulting DNA fragments.

both in relation to the recognition sequence itself and in relation to each other, are determined by the identity of the restriction enzyme used to cleave the DNA molecule in the first place. Once the enzyme cuts have been made, the original molecule breaks into fragments. The number of fragments produced depends on the number of times that a specific recognition sequence appears along the length of the molecule. The characteristics of the newly produced fragment ends, often called **termini**, are also determined by the identity of the restriction enzyme which created them.

SOME RESTRICTION ENZYMES CREATE BLUNT ENDS As is shown in Figure 4.11, the restriction enzyme *Hae*III, isolated from the bacteria ***Haemophilus aegyptius***, has a recognition sequence of GGCC. *Hae*III cleaves both strands of the DNA between the G and the C residues. As a result, the two new fragments have square or blunt ends.

SOME RESTRICTION ENZYMES CREATE STAGGERED ENDS The restriction enzyme *Eco*RI recognizes the base sequence GAATTC. Unlike *Hae*III, *Eco*RI does not cleave its recognition sequence in a blunt-ended fashion. Instead, *Eco*RI cleaves between the G and the A residues on both DNA strands. Once the cuts have been made in the DNA backbones, the resulting fragments are held together only by the relatively weak hydrogen bonds which hold complementary base pairs together in the DNA double helix. Since these bonds are not strong enough to hold the fragments together for long at 37°C (98°F), the fragments soon separate from one another, leaving behind protruding 5' ends composed of unpaired bases.

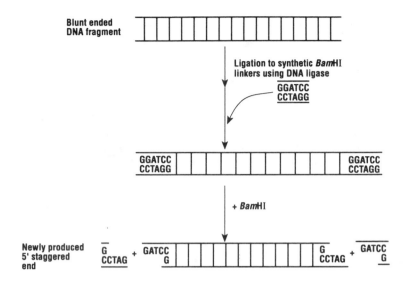

Figure 4.12
An example of the use of linkers. *Bam*HI linkers can be used to convert a blunt-ended DNA fragment into a fragment with protruding 5′ ends. Linkers containing the *Bam*HI recognition sequence are ligated to the blunt ends of a DNA molecule. The linkers are then cleaved with *Bam*HI to produce the protruding 5′ ends typical of this enzyme.

Other enzymes, for example **HbaI,** work in the reverse fashion. This enzyme, isolated from the bacteria ***Haemophilus haemolyticus,*** has a recognition sequence of GCGC. *Hba*I cleaves the DNA backbones on both strands between the G and C residues closest to the 3′ end of the recognition sequence. As a result, when the fragments separate from each other, protruding 3′ ends remain.

Linkers convert Blunt Ends into Staggered Ends

Linkers, chemically synthesized lengths of nucleotides, contain a specific recognition sequence often used to convert a blunt-ended DNA fragment into one with staggered ends. This conversion process involves attaching a linker to the blunt end of a DNA fragment using the enzyme DNA ligase (see below). The linker-fragment complex can then be treated with the restriction enzyme whose recognition sequence is contained in the linker. The resulting cleavage of the linker converts the formerly blunt end into a staggered end as described in Figure 4.12.

Staggered Ends Help Form Recombinant DNA Molecules

The staggered cuts which are made by a variety of restriction enzymes, including *Eco*RI and *Hba*I as mentioned above, invaluably aid in the various processes of genetic engineering. As we have seen, the tails which remain after staggered cuts are made are composed of short lengths of unpaired bases. Scientists sometimes call these tails "sticky ends" due to their complementarity with one another.

In order for either 3' or 5' tails to be sticky, there must be complementary sequences of bases on the two fragment tails in question. For instance, the protruding ends generated by the enzyme *Bam*HI will not bind with the protruding ends generated by the enzyme *Eco*RI, because the *Bam*HI tails have the sequence GATC while the *Eco*RI tails will have the sequence AATT. The *Bam*HI sequence does not complement the *Eco*RI sequence. Thus, these tails are not sticky with each other.

On the other hand, any two DNA fragments produced by the *same* enzyme will stick together due to the complementarity of their sticky ends. For example, cleavage with *Bam*HI produces a 3' tail whose sequence is "...CTAG" along with a tail whose sequence is "GATC...." These two tails complement one another, and are thus true sticky ends with each other.

The bonding of sticky ends occurs between *any* two sequences of unpaired bases, regardless of the sources of the DNA molecules, as long as the two involved tails have complementary base sequences. This means that fragments of DNA from any two cells, be those cells as unrelated as cabbage and cow cells or cauliflower and collie pup cells, can be joined together by the complementarity of their sticky ends to form a new molecule. Such a molecule, which consists of pieces of unrelated DNA which have been joined together, is called a **recombinant DNA molecule.**

Figure 4.13 shows how a recombinant DNA molecule might be constructed from a length of human DNA and a length of bacterial DNA if they were both digested with the enzyme *Eco*RI which cleaves between the G and A residues in its recognition sequence of GAATTC. All of the DNA fragments resulting from this digestion will have protruding 5' ends with the base sequence AATT. When these human and bacterial DNA fragments are mixed, base pairing occurs between the complementary unpaired bases of the protruding ends. The fragments then fuse via the stickiness of their sticky ends, forming new molecules. Because any two complementary sequences can bind to each other, some of the newly formed molecules will consist of two or more bacterial fragments while some will consist of two or more human fragments. Only some of the newly fused molecules will consist of both human and bacterial fragments. The term "recombinant DNA" can only properly be applied to a molecule which is composed of fragments that were not originally adjacent to one another.

DNA Ligase Completes the Recombinant DNA Molecule Although sticky ends can be held together for a short time by the hydrogen bonding between the complementary bases of the protruding ends, these types of bonds are not strong enough to keep the ends together indefinitely at physiological temperatures. To make the molecule stronger, the enzyme **DNA ligase** is added to the preparation. This enzyme, which is isolated from a bacteriophage known as T4, seals, or **ligates,** the nicks in the DNA backbones where the DNA fragments come together. DNA ligase acts by forming a new chemical bond between any adjacent nucleotides having free 5'-phosphate and 3'-hydroxyl groups (see Figure 1.8). Once DNA ligase makes these new, strong, covalent bonds between the DNA fragments, the new molecule is complete. Many of these new molecules have

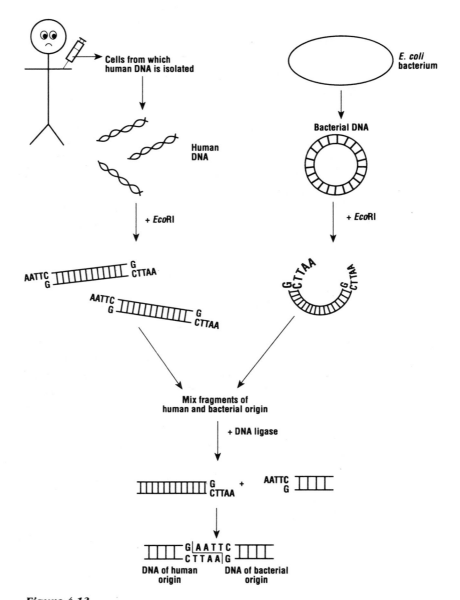

Figure 4.13
DNA fragments—regardless of the species of origin—can be ligated to other DNA fragments of any species via the complementarity of their "sticky ends."

some extremely unusual properties—they are composed of pieces of DNA that initially were not adjacent to one another. Indeed, they may very well have started out in the nuclei of entirely different organisms!

REVIEW AND SUMMARY

1. All organisms contain detailed hereditary information in the form of **nucleic acids,** which are composed of subunits known as **nucleotides.** In most cases these nucleic acids are DNA molecules which are folded into precise patterns, around "spools" of histone proteins, to form **chromosomes.**

2. Cells, the fundamental unit of all living organisms, are divided into two categories: **eukaryotic** cells which have nuclei, and **prokaryotic** cells which lack nuclei.

3. Eukaryotic cells generally have multiple, linear chromosomes while prokaryotic cells generally have single, circular chromosomes.

4. In addition, prokaryotic cells often contain **plasmids,** circular DNA molecules separate from the chromosome. Plasmids frequently contain genes that allow a cell to survive in the presence of substances, such as antibiotics, which are normally toxic to that cell.

5. Both plasmid and chromosomal DNA molecules can be isolated from cells. In both cases, the cell membrane is solubilized by the application of detergent. The resultant lysate is then enzyme-and/or heat-treated to remove various contaminants from the desired DNA. The DNA molecules are collected by precipitating them into ethanol.

6. **Restriction enzymes,** found in bacterial cells, function to protect those cells from infection by bacteriophage particles. They carry out this function by cleaving the invading phage DNA. DNA of the bacterial host cell remains intact because it is protected by the presence of methyl groups which are attached to some of the bases of the bacterial DNA. Invading phage DNA lacks these methyl groups at nuclease recognition sites and is thus subject to cleavage by the restriction enzymes.

7. Restriction enzymes function by binding to specific sequences of bases known as **recognition sequences.** Each enzyme has a single specific recognition sequence. Once the enzyme has bound to the specific base sequence, it cleaves the DNA backbone, thus breaking the molecule into fragments. The most useful enzymes, known as **type II restriction endonucleases,** cleave the DNA molecule at a predictable site within the recognition sequence itself.

8. Each different restriction enzyme leaves characteristic ends—**termini**—on the fragments which result from restriction cleavage. Two different types of termini are produced: 1) "blunt ends" in which there is no overhanging single-strand tail remaining after cleavage, and 2) "staggered ends" in which a single-strand tail remains. The tail can extend in either the 3' or the 5' direction.

9. The staggered ends resulting from cleavage by restriction enzymes complement one another in certain situations. These complementary staggered ends tend to hydrogen bond to one another and are thus called **"sticky ends."**

10. Sticky ends make it possible to "glue" two DNA fragments together, regardless of the sources of the DNA. The enzyme DNA ligase functions to complete the sugar-phosphate backbone between the newly joined or ligated fragments. When the fragments come from at least two different sources, the resultant ligated molecule is called **recombinant DNA.**

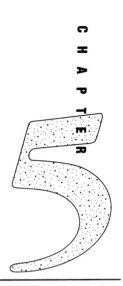

Preparing a portrait: The characterization of a DNA molecule

Time: The Present
Place: Metropolitan Hospital

Twenty-two-year-old Kelly S. sits in front of Dr. Malone's desk. Kelly's mother, age fifty-two, recently died from Huntington's disease. Kelly knows that a genetic mutation causes this invariably fatal degeneration of the central nervous system. She also knows that there is a fifty percent chance that she has inherited that mutant gene from her mother. Kelly is concerned about the future: will she also eventually succumb to the disease?

Dr. Malone tells Kelly that a certain test can tell her, with ninety-six percent accuracy, if she has the gene for Huntington's disease. The test requires blood samples from many members of Kelly's extended family. "But," Dr. Malone cautions Kelly, "think long and hard about it. Do you really want to know? After all, even though we can currently examine certain characteristics of your chromosomes to determine whether or not they are associated with the Huntington's gene, we cannot—given today's medical knowledge—correct the gene if you have it."

Time: The Future
Place: Metropolitan Hospital

Newlyweds Michael and Susan W., seated in front of Dr. Dean's desk, appear concerned and upset. Dr. Dean just told them that he found some problems in the routine chromosomal characterization required of all newly married couples. Dr. Dean continues to speak: "Michael and Susan, as you know, we examined your chromosomes so that we could tell you if there were likely to be any genetic problems with the children that you might someday have. When we examined your chromosomes we defined all of the genes which make each of you into a unique individual. Unfortunately, both of you carry a rare, recessive gene which affects development of the fetal nervous system during pregnancy. Many years ago, young couples in your position would have had no choice but to play the odds if they wanted children. There was a one-in-four chance with each and every pregnancy that the developing fetus would inherit a bad gene from both parents. If that happened, the pregnancy ended in miscarriage because of major spinal and neural tube defects in the fetus.

"The two of you are luckier than our grandparents were. Due to advances in biological and medical sciences, we can take a sample of DNA from each of you and characterize it very precisely. Since we know what every one of the approximately 100,000 genes that make up a human being should look like, we can find the recessive gene that may cause you problems. Once we've found the problematic gene, we use special enzymes called restriction endonucleases to remove that gene from samples of your eggs, Susan, and Michael's sperm. We'll order replacement copies of the gene from RepGeneSys, and we'll fix the genomes of your germ cells by making recombinant DNA molecules out of fragments of your DNA and the replacement gene. Then, we'll allow the recombinant sperm to fertilize a recombinant egg, implant the embryo in your uterus, and you should have a fine, healthy baby!"

With the limited biological and medical knowledge of today, this scene of the future is yet fiction. Although scientists have found the approximate locations of about 1,500 genes in the human genome, more than 98,000 still remain to be located. In addition, of the approximately 100,000 genes that make up a human being, few actual base sequences have been determined. So today, even if we are aware that we are carrying a bad gene, we can't do much about it. Each day, however, research progresses and we come closer and closer to a precise characterization of the human genome. Someday, in the not too distant future, we will know not only the chromosomal location of every human gene, but also the base pair sequence of every gene in the human genome. With this knowledge, micromanipulations of the genome will become increasingly possible, and the scene described above may become more than science fiction.

OVERVIEW

All living organisms reflect a complex series of plans and instructions, found in the genetic material of each individual organism. We know that each individual has a unique series of genetic instructions, and that to change those instructions means to change the individual itself.

By studying the hereditary material of a variety of organisms, scientists have found many ways to manipulate the microscopic molecules of DNA. Some of these manipulations—including DNA isolation, cleavage and ligation briefly described in chapter four—have allowed scientists to develop ways of characterizing specific, individual DNA molecules. This chapter examines some of these characterization methods and discusses ways in which future scientists might characterize DNA. It also answers the following questions:

1. How do restriction endonucleases help us understand the structure of genetic material?

2. What role does an electrical field play in the analysis of the base sequence of any DNA molecule?

3. What do we know about the actual base sequence of the human genome?

MAKING AND SEPARATING FRAGMENTS OF DNA MOLECULES

Restriction Endonucleases Cut DNA in a Reproducible Fashion

We have already learned that any specific fragment of DNA differs from any other specific fragment of DNA in a fundamental way: each fragment has its own unique sequence of base pairs, and different fragments always have different sequences of base pairs. In fact, we can identify a particular length of DNA by its base sequence just as we identify our friends and family based on physical appearance and other individual characteristics. If we see someone who looks just like our best friend, except that the new person is two inches taller and blond rather than brunette, we know that that person is not actually our best friend, but just resembles her. Likewise, when we look for the piece of DNA responsible for making insulin—the insulin gene—we recognize that length of DNA based on its physical characteristics, or, more specifically, its base sequence. If we find a piece of DNA with *almost* the same base sequence as the insulin gene, we know that although the new DNA is similar, it is not actually the insulin gene. In other words, every copy of a normal insulin gene has the same base sequence as any other copy. Put more universally: genes are defined by, and can therefore be identified by, their base sequence.

Each restriction endonuclease has a particular recognition sequence of base pairs, and a restriction endonuclease only cleaves DNA when it encounters its own specific recognition sequence (Chapter Four).

Knowing these two facts:

1. Each individual gene has the same base sequence in every cell of an organism, and in all organisms of the same species, as long as there are no mutations involved.

2. Restriction endonucleases cleave DNA only in the vicinity of a specific and invariable recognition sequence of base pairs.

We can draw an important conclusion: any particular restriction endonuclease will cleave a specific DNA molecule in the same places no matter how many times we mix samples of the enzyme with copies of the DNA molecule.

For example: let's say that we have a molecule of DNA that is 1,000 deoxyribonucleotide pairs in length. Let's also say that the base sequence GGATCC occurs only once along the entire length of the molecule, and that it occurs so that the first guanine residue in the sequence is base #297 of the 1,000 bases which make up the DNA molecule, and the last cytosine residue is base #302. If we mix this molecule of DNA with the restriction enzyme *Bam*HI, whose recognition sequence just happens to be GGATCC, it cleaves the DNA molecule in exactly one place. As a result of this enzymatic cleavage, the molecule which started out as a single fragment composed of 1,000 base pairs becomes two fragments, one of which will be approximately 300 deoxyribonucleotide pairs long, and the other of which will be 700 deoxyribonucleotide pairs long (Figure 5.1).

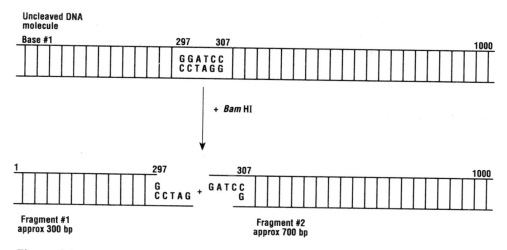

Figure 5.1
*Bam*HI cleavage of a sample DNA molecule results in the production of two DNA fragments. One fragment approximates 300 base pairs in length while the other is about 700 base pairs in length.

Anytime—and indeed everytime—we mix another set of copies of our 1000-base-pair DNA sequence with the enzyme *Bam*HI, the same thing happens. In fact, if we mix a piece of unidentified DNA with *Bam*HI and the enzymatic reaction produces a 300-base-pair fragment along with a 700-base-pair fragment, the unknown DNA might be a copy of the original DNA molecule. To make a positive identification we would examine both molecules—the identified as well as the unidentified—with a variety of restriction endonucleases. If the same cleavage results occurred when each of the molecules was treated with each of the enzymes, we could postulate that the two molecules were, in fact, copies of one another.

ELECTROPHORESIS CAN SEPARATE DNA FRAGMENTS FROM ONE ANOTHER

When DNA molecules are treated with restriction endonucleases, the various fragments that are generated can be separated from one another using a relatively simple technique called **agarose gel electrophoresis.** In this technique, an electrical field is used to separate DNA fragments based on both their electrical charges and relative sizes. Electrophoresis gets its name from the combination of the prefix "electro" and the Greek word "phoresis," meaning "to be carried." In other words, the technique of electrophoresis means that something is "to be carried by electricity."

Agarose Acts as a Molecular Sieve

Agar, a gelatinous material isolated from various seaweeds, is used in the foods of many cultures as well as in various scientific endeavors. If you have ever had the doctor take cells from your throat to be cultured to detect a possible strep infection, you have benefitted from the use of agar (Figure 5.2). Agarose is a

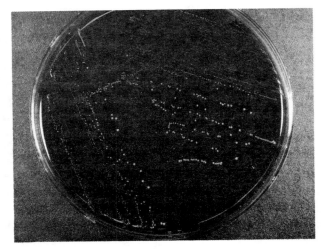

Figure 5.2
(a) A doctor swabs this patient's throat to check for the presence of *Streptococcus* bacteria. (b) Material from the throat collected in the swabbing procedure is placed on a layer of blood agar—agar mixed with intact red blood cells. When *Streptococcus* bacteria are present, their growth causes clear spots to appear in the blood agar. The clear spots result from the hemolytic activity—the ability to lyse red blood cells—characteristic of the *Streptococcus* bacteria.

purified powder isolated from agar. Agarose powder, mixed with water and then boiled, forms a solution that will eventually solidify into a gel, much like a gelatin dessert.

As the agarose gel solidifies, the molecules of agarose line up in a dense three-dimensional pattern. The pattern contains holes or spaces, called **pores,** between the actual agarose molecules. Because of the precise way in which the agarose molecules line up to produce a pattern of pores, a molecular sieve forms as the agarose solidifies (Figure 5.3). This sieve allows some molecules to slide right through the pores, but causes other molecules to proceed more slowly. The variation in the speed with which different molecules move through the pores of the sieve depends on the electrical characteristics of those molecules, as well

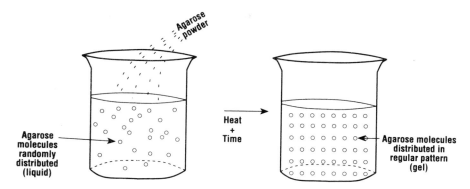

Figure 5.3
The agarose molecules in a liquid suspension are randomly distributed. After being boiled and cooled the agarose molecules align in a more regular grid pattern and the suspension changes from a liquid to a gel.

as on the relative sizes of the molecules and of the pores of the sieve. The size of the pores in the sieve can be varied by changing the amount of agarose powder that is mixed with the water to form the gel: a higher percentage (e.g. 2%) of agarose powder yields smaller pores in the finished gel; this efficiently separates small fragments from one another. A lower percentage (e.g. 0.3%) of powder produces larger pores which more efficiently separate larger fragments. Appropriate manipulation of the percentage of agarose in the gel allows separation of linear DNA fragments ranging in size from approximately 100 nucleotides to approximately 60,000 nucleotides.

DNA Fragments Are Separated Using an Electrical Current

To separate DNA fragments using the technique of agarose gel electrophoresis, scientists suspend a **comb** so that its teeth are partially submerged in the molten agarose which fills a rectangular mold or **casting tray.** After cooling, the resulting slab of solidified agarose is submerged in a mixture of salts and water. The comb is carefully removed, leaving behind empty wells which are filled with a DNA sample consisting of DNA fragments and a **loading dye** made up of sucrose and one or more visible dyes. The relatively dense sucrose serves to sink the DNA sample into the well so that it doesn't float away, while the dye visually marks the electrophoretic progress of the invisible DNA fragments through the gel. An electric field is then set up in the gel and the surrounding salt water by applying an electric current to electrodes set up at opposite ends of the gel slab (Figure 5.4). Two important facts now come into play in the process of gel electrophoresis:

1. DNA molecules have a negative electric charge due to the phosphate groups which alternate with sugar molecules to make up the backbone of the DNA double helix.

    and

2. Opposite electric charges tend to attract one another.

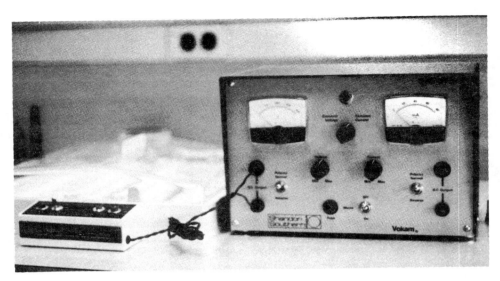

Figure 5.4
The basic equipment for agarose gel electrophoresis includes an **electrophoresis box** which has both a positive and a negative electrode, a shelf designed to hold the **gel,** a **comb** used to form the wells within the gel, and a power supply capable of creating an electric field within the salt solution in the box.

Taken together, these facts mean that any DNA molecules in the agarose gel move or **migrate** through the gel toward the positive pole of the electric field with a speed that depends on the relative sizes of the DNA molecules. Larger molecules move more slowly than smaller molecules. Generally speaking, molecules of the same length will move the same distance into the gel under the same conditions of time, temperature, and voltage.

A Population of DNA Can Be Isolated Based on Size

Using agarose-gel electrophoresis, a researcher can separate a mixed assortment of DNA molecules, which differ from one another by their length, into separate, discrete populations of uniformly sized molecules. Since all DNA molecules of a given size will migrate approximately the same distance into the gel, they tend to form a **band** at some particular location within the gel. Each band represents many, many copies of DNA fragments, all of which are approximately the same length. DNA molecules which differ in length by at least thirty to fifty nucleotides will form bands in different locations. The size of a DNA fragment can be determined by electrophoresing a DNA sample of **reference fragments**—fragments whose sizes are known—along with the fragments of unknown length. In this way the band positions of the unknown fragments can be compared to the band positions of the reference fragments. Since band position is directly related to fragment size, with the distance of the band from the original well being inversely proportional to the size of the fragment, such comparisons should provide the desired information.

Neither DNA molecules themselves, nor the bands produced by electrophoresis, are visible to the unaided human eye. Therefore, in order to determine

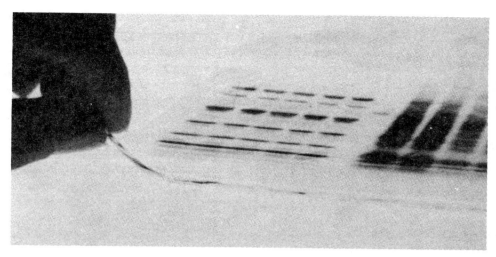

Figure 5.5
Gel electrophoresis allows separation of molecular fragments into homogeneous bands based on size and charge differences. Fragments to be separated are loaded into wells at one end of the gel. An electric current—generated between a negative electrode at the top or loading end of the gel, and a positive electrode at the bottom end of the gel—then causes the fragments to move through the pores of the agarose gel. In general, smaller fragments move farther into the gel than larger ones with the distance of migration being inversely proportional to the fragment size. Fragment size can be estimated by comparing the distance that a fragment of unknown size migrates with the distance traveled by a fragment of known size.

the location of the DNA bands which result from the process of electrophoresis, the gel is treated with a stain which binds preferentially to the DNA molecule and leaves the gel itself unstained—the bands take on the color of the stain while the gel remains relatively clear or milky opaque (Figure 5.5). Neither the process of electrophoresis, nor of staining, need irreversibly harm the DNA molecule. This means that a band—representing a population of size-purified DNA—can literally be cut out of the gel, and the DNA can be electrophoresed out of the band and into a surrounding salt solution. This size-purified DNA can then be used in a variety of ways.

IN DETAIL: AGAROSE GEL ELECTROPHORESIS

Separation of DNA fragments of differing lengths depends on the ability of agarose to form a **molecular sieve,** a sort of strainer or colander that allows smaller molecules to pass through more quickly than larger molecules. The molecular sieve is produced by adding varying amounts of agarose to a mixture of salt and water, which is a good conductor of electricity. Agarose gels generally range in concentration from 0.8% (all solution concentrations in this text are given in weight/volume or w/v terms. Thus the percentage is the number of grams of chemical/100 ml of liquid. For example, an 0.8% agarose gel consists of 0.8 grams of agarose for every 100 ml of saltwater) to 2.0%, with the lower concentrations producing softer gels that allow larger molecules to pass through. Conversely, the higher concentrations produce harder gels and allow only the relatively smaller molecules to pass. Scientists choose gel concentrations appropriate to the sizes of fragments which they hope to separate.

Agarose gels are produced by adding the appropriate amount of agarose powder to a salt solution, and heating the mixture to dissolve the agarose. The molten agarose is poured into a **casting tray** which resembles a shallow, flat, rectangular box. A **comb** is positioned so that its wide, flat teeth extend into the molten agarose at one end of the casting tray but do not touch the tray itself. When the gel cools and solidifies, it is then immersed into a salt solution which will conduct electricity. The comb is then carefully removed. It leaves behind a series of holes called **wells** in the newly formed gel.

The DNA to be electrophoresed is digested with restriction endonucleases to produce a series of fragments of varying lengths. The digested DNA is mixed with sucrose and a dye. Together, these components, known as **loading dye,** allow the DNA to be placed into the wells of the gel: the sucrose increases the density of the DNA preparation so that it sinks to the bottom of the well, while the dye increases the visibility of the preparation. At least one well of each gel will be filled with **reference DNA**—fragments of known lengths—establishing a known scale against which fragments of unknown lengths can be compared.

Electrical current is then applied to the poles at opposite ends of the electrophoresis chamber. The current can be varied according to the specific needs of the researcher; typical values range from 50 to 150 volts. Greater voltages result in faster migration of the DNA in the gel, however, increased voltages also produce heat which can, in some cases, actually melt the agarose gel! The DNA-loading dye mixture will migrate out of the wells and into the gel in response to the electrical field created by the applied current. Progress of the DNA into the gel can often be monitored by watching the **dye front,** or the position of the loading dye, as it progresses through the gel: dye molecules, generally extremely small, migrate ahead of most DNA fragments. Electrophoresis is halted when the dye front moves two-thirds to three-fourths of the way into the gel. If electrophoresis continues too long, some, if not all, of the DNA fragments will actually migrate off the end of the gel to be lost in the salt solution in which the gel is submerged.

After electrophoresis, the gel is removed from the electrophoresis chamber and stained to make the DNA bands more easily seen. The two dyes most commonly used for this purpose are called **ethidium bromide** and **methylene blue.**

Ethidium bromide: this chemical, a potent mutagen, must be handled with extreme care! A gel is soaked in a one ug/ml solution of ethidium bromide for five to ten minutes. The chemical quickly distributes itself throughout the gel and concentrates wherever DNA molecules are present. In fact, the ethidium bromide molecules actually insert themselves—**intercalate**—between the bases of the DNA molecule. When the gel is rinsed to remove excess ethidium bromide and is exposed to ultraviolet light, the DNA appears as fluorescent orange bands.

Methylene blue: researchers sometimes use this chemical in place of ethidium bromide because it is less dangerous as a mutagen. However, it is also significantly less sensitive than ethidium bromide, and slower to stain the electrophoresed DNA. To use this chemical, a gel is soaked in an 0.025% solution of methylene blue for 15 to 30 minutes. The best staining results are obtained when the gel is then **destained** by soaking it in plain distilled water for several hours. Methylene blue causes the DNA bands to appear blue under normal room illumination.

The distance that a DNA fragment migrates into the gel is inversely proportional to its molecular weight or number of base pairs. In other words, the smaller the fragment, the farther it migrates.

CHARACTERIZATION OF DNA FRAGMENTS

DNA Can Be Mapped Using Restriction Endonucleases

Its own unique sequence of deoxyribonucleotide pairs characterizes each and every molecule of DNA and we know that restriction enzymes cleave DNA only in the vicinity of particular sequences of base pairs known as recognition sequences. This means that the cleavage sites on a DNA molecule are constant: every time DNA molecule A is treated with enzyme B—under the same conditions of time, concentration, and temperature—the same fragments will be produced. Therefore, DNA molecules can be characterized by the sizes of the fragments which are produced as a result of cleavage by any particular enzyme. As we have seen, these fragments can be separated and their sizes determined by agarose gel electrophoresis. The band patterns generated by this procedure can, in turn, be used to prepare **restriction maps** for different DNA molecules. Restriction maps, schematic representations of DNA molecules, show the locations of recognition sequences for various restriction enzymes.

Daniel Nathans, working at Johns Hopkins University in Baltimore, Maryland, produced the first restriction map in 1971. Dr. Nathans, using the endonuclease *Hin*dII, cut the circular chromosome of the virus known as SV40. He saw that a very short superficial treatment of the SV40 DNA with *Hin*dII resulted in a single cut in the viral DNA. The single cut, which resulted from the incomplete enzyme digestion, transformed the circular chromosome into a linear one. As the enzyme was allowed to interact with the SV40 DNA for increasing amounts of time, it produced increasing numbers of cuts in the molecule until a maximum of 11 fragments were produced. By using gel electrophoresis, Dr. Nathans determined the sizes of the 11 fragments. After looking at the pattern of fragments generated by the incomplete digestions and comparing them with the fragments of the complete digestion, he then drew a map of the circular SV40 chromosome showing the location of those 11 *Hin*dII recognition sequences. This analysis yielded the first restriction map.

Restriction maps containing greater detail can be produced by digesting the DNA molecule to be mapped with a variety of restriction enzymes. Accurate restriction maps are invaluable because they provide a detailed description of any areas of the DNA molecule which are susceptible to cleavage by a restriction enzyme. Restriction maps serve a function similar to a good road map: although you might find your destination in an unfamiliar town without a street map, an easier and more direct trip ensues if you plan it with an accurate map. Similarly, you can stumble upon the restriction sites which will allow you to remove a particular segment from the middle of a DNA molecule by yourself, but access to an accurate restriction map markedly simplifies your task.

DNA Is Polymorphic

Let's say that you and your neighbor are both the proud owners of shiny, new ten-speed bicycles. Both bicycles have the same basic components: handlebars, caliper brakes, racing seat, diamond-shaped frame, and so forth. Both bicycles also have the same basic structure: handlebars in front, seat in the middle, pedals on the bottom, and so forth. Despite these similarities, however, the bicycles are not identical: your red bicycle has racing handlebars, while your friend's bicycle is blue with all-terrain handlebars. Although the bicycles are basically the same, some differences do exist.

Genes resemble these bicycles in that any two members of the same species have the same basic arrangement of genes on their chromosomes. Furthermore, the base sequence of any particular gene in different members of the same species is basically the same. For example, you and your neighbor both have a gene on chromosome 22 that plays a role in the way that your immune system fights off disease and illness. Both copies of this gene—yours and your neighbors'—have the same basic structure and work in the same basic fashion. However, most likely the actual base sequence of these two genes differs slightly. This slight difference provides an example of a **genetic polymorphism,** a variant in the base pair sequence of either the coding or the non-coding portion of a DNA molecule that occurs with a frequency of one percent or more. A polymorphism in the coding sequence of a gene may lead to differences in the gene's polypeptide product, such as the difference between normal and sickle-cell hemoglobin. Non-coding polymorphisms, which occur at approximately 1 nucleotide in every length of 250 nucleotides, do not appear to have any phenotypic effects. Despite their lack of apparent effect on the function of a gene, non-coding polymorphisms can be used as genetic markers.

DNA Polymorphisms Can Be Used as Genetic Markers

The genes of eukaryotic organisms are separated from one another along the length of a chromosome by **non-coding sequences** of bases. Although these sequences do not appear to be functional genes or to encode polypeptide products, they do appear to be **conserved.** That is, any particular sequence of bases is very similar when compared among all members of a given species. However, these non-coding sequences are also prone to the slight variations in base sequence that we call polymorphisms. Although these polymorphisms do not affect the overall functioning or general appearance of the organism, they can be significant. For example, humans have approximately one polymorphism in every length of 250 non-coding base pairs. This level of genetic polymorphism means that genetic variation can occur at a great number of locations without harming the overall functioning of the organism. The identity of the bases involved in the genetic polymorphisms draws a portrait of an individual's genome that is as unique as a fingerprint.

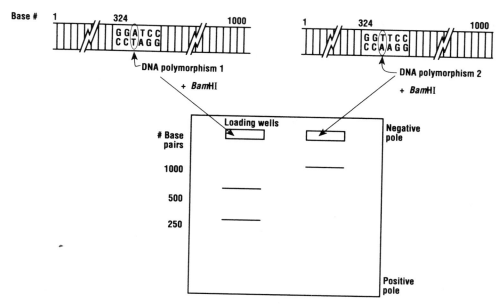

Figure 5.6
Restriction Fragment Length Polymorphisms (RFLPs) can be detected by cleavage with appropriate restriction endonucleases and subsequent gel electrophoresis.

DNA Polymorphisms Yield RFLPs

The presence of a genetic polymorphism can often be detected by using a variety of restriction endonucleases. These useful enzymes cleave a molecule of DNA only after binding to particular recognition sequences. Each enzyme has its own recognition sequence, and that sequence of bases never varies. If even a single base changes in the recognition sequence, it no longer functions to activate the cleavage activity of the enzyme. This means that the restriction endonuclease will no longer cut the DNA molecule.

This remarkable enzyme specificity can be exploited to detect genetic polymorphisms. Let's say that we have two preparations of DNA molecules with identical base sequences, 1,000 base pairs in length, except for the following polymorphism: DNA #1 has the sequence GGATCC at a location beginning with base #324, while DNA #2 has the sequence GGTTCC in the same location. What happens when we treat each of these DNA preparations with the restriction endonuclease *Bam*HI—an enzyme whose recognition sequence happens to be GGATCC? Gel electrophoresis of DNA fragments resulting from such an experiment shows a major difference. Treatment of DNA molecule #1 with *Bam*HI will yields two bands on the gel, while treatment of DNA #2 yields only a single band (Figure 5.6). *Bam*HI cannot cleave DNA #2 because the genetic polymorphism has destroyed its recognition sequence. Thus, major band-pattern differences result from very small genetic polymorphisms. The variable fragment lengths resulting from genetic polymorphisms are called **Restriction Fragment Length Polymorphisms** or **RFLPs**.

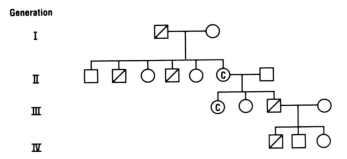

Figure 5.7
Pedigree for four generations of a family with color-blindness. Squares = normal males. Slashed squares = affected males. Circles = normal females. Circled Cs = carrier females. Carriers, individuals who possess a gene for a particular trait, do not exhibit the phenotype of that trait. Carriers are able to pass the trait in question on to their offspring.

RFLPs Can Help to Predict or Diagnose Disease

We know that physical traits are passed from generation to generation among the members of a family. The pedigree in Figure 5.7 shows how the phenotypic trait of color blindness can be traced among generations of a single family. To trace the passage of a particular phenotype in your family, look for traits such as the presence or absence of a widow's peak or hitchhiker's thumb, or examine the distribution of blood type or eye color in your extended family. If you had access to the appropriate laboratory tools you could also look at the distribution of RFLPs in your family. Because RFLPs are the direct result of DNA polymorphisms and since accurate copies of DNA molecules are included in the germ cells which grow into the next generation, RFLPs themselves pass from parent to offspring in much the same way as eye color or blood type.

Extensive analysis of the RFLPs present in some extended families shows that some RFLPs signal the probable presence of certain diseases as well. For example, in 1985 scientists James Gusella, Nancy Wexler, and Michael Conneally completed a study in which they determined the identities of various RFLPs that were present in more than five hundred members of one extended family living in both the United States and Venezuela. This family was unique in that many of them suffered from Huntington's disease. Gusella, Wexler, and Conneally discovered that those members of the family who were affected with the disease also had a genetic polymorphism which resulted in a RFLP that they named G8. Members of the family who did not have Huntington's disease did not have RFLP G8.

RFLPs, present in the germ cells that come together to form a person, can be detected at any time during the life span of that person. Huntington's disease, on the other hand, does not develop in most individuals until middle age. Because scientists have not yet identified the actual gene or genes that cause Huntington's disease, it has not been possible to know whether an individual had inherited the disease until its actual onset. The studies described above raise the possibility of predicting the possible future onset of Huntington's disease in an individual. Scientists feel that the presence of G8 indicates a ninety-six percent probability of the future development of Huntington's disease, thus making RFLP G8 a valuable diagnostic tool.

THE BASE SEQUENCE OF DNA FRAGMENTS CAN BE DETERMINED

Although restriction maps and RFLPs provide valuable information regarding the base sequence of a DNA molecule, the information is woefully incomplete. To make the most efficient and effective manipulations of a length of DNA, a scientist frequently needs information regarding the complete base sequence of the DNA molecule. Of the variety of ways to get this information, all of the techniques rely on the fact that DNA fragments can be separated on the basis of length by gel electrophoresis. We have already discussed agarose gels, which separate fragments of DNA differing from one another by at least thirty to fifty bases in length. In DNA sequencing, however, it is necessary to separate DNA fragments differing in length by only a single base. This is accomplished by **denaturing** double-stranded DNA molecules so that the two strands separate from one another and electrophoresing the single strands on a type of gel known as a **polyacrylamide gel.** Such gels separate DNA molecules that are between 15 and 600 deoxyribonucleotides in length at the resolution of a single base. That is, this type of gel separates a fragment of DNA composed of 451 deoxyribonucleotides from one composed of 452 deoxyribonucleotides.

Maxam and Gilbert DNA Sequencing

Scientists Allan Maxam and Walter Gilbert, working at Harvard University in 1977, developed a method for determining the sequence of bases in a DNA molecule. The Maxam and Gilbert method depends on the ability of certain chemicals to selectively degrade particular bases within a molecule. This technique has five basic steps:

1. One end of each strand of the DNA molecule to be sequenced is radioactively labelled and the strands are separated from one another.

2. A particular base, or pair of bases, is chemically modified—for example some of the guanine bases or some of the cytosine bases are modified.

3. The modified bases are removed from the DNA backbone.

4. The backbone is cleaved wherever a base is missing.

5. The resulting single stranded DNA fragments are size-separated by polyacrylamide gel electrophoresis.

Let's look at the steps in a little more detail (Figure 5.8). Once many copies of the soon-to-be-sequenced DNA molecule have been radioactively labelled at one end, they are denatured and divided into four samples or **aliquots.** Various chemical treatment of each of the four aliquots causes modification in 1) the guanine bases, 2) both adenine and guanine bases, 3) the cytosine bases, and 4) both thymidine and cytosine bases, respectively. The chemical reaction in

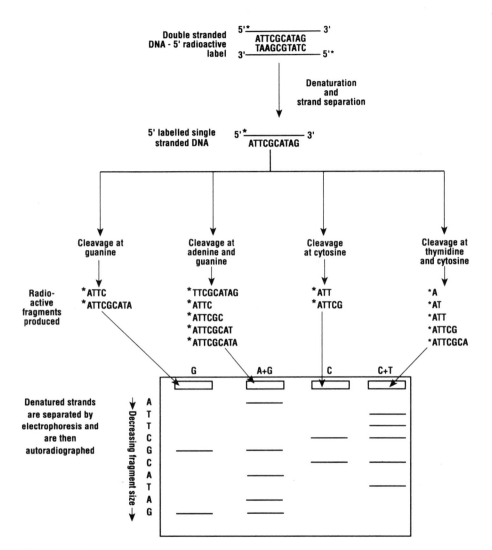

Figure 5.8
Maxam and Gilbert technique for DNA sequence determination. A double-stranded DNA molecule, radioactively labelled at the 5′ ends of each strand, is denatured and a preparation is made of one of the labelled strands. The strand preparation to be mapped is divided into four aliquots, each of which is chemically treated to cause modification and removal of one or two specific bases. The reactions are carried out so that an average of only one base is modified and removed from each individual molecule in an aliquot. The DNA backbone is cleaved at the site of each missing base. The resulting fragments are then size-separated by gel electrophoresis.
Autoradiography indicates the position of all 5′-labelled fragments. The length of the fragments directly indicates the position of the modified bases. By relating the size of any individual fragment with the identity of the modified base or bases, the precise position of all bases in the original DNA fragment can be determined. Thus, the base sequence of the mapped fragment can be read directly from the autoradiograph.

each aliquot is controlled so that it remains incomplete. This means that only a few target bases are modified in any particular DNA molecule. Since each aliquot contains many DNA molecules, all of the target bases are modified somewhere within the whole population of molecules that make up each aliquot. All of the base-modified aliquots are then treated with a second chemical which causes the DNA backbone to break wherever there is a modified base. This breakage results in the production of a series of DNA fragments of varying lengths, some of which are derived from sequences located between one end of the molecule and an interior break. Other fragments represent lengths of DNA found between two base-modified sites. Only the first type of fragments—those extending from one end of the molecule to a base-modified site—will be labelled.

All fragments, regardless of the presence or absence of the radioactive label, separate on the basis of size by electrophoresis on a polyacrylamide gel. Band patterns form on the gel in the same way that band patterns form on an agarose gel. However, rather than staining the gel to see the band pattern, we will photograph it. When the gel is placed on undeveloped photographic film, the many radioactively labelled DNA fragments in each band expose it. Thus, the exposed film shows the band pattern of the polyacrylamide gel. This technique, in which radioactive molecules are detected by their ability to expose photographic film, is called **autoradiography.**

We know that polyacrylamide gels resolve DNA fragments which differ from one another by only a single base in length. Therefore, we can read the base sequence of our DNA molecule directly from the band pattern on the exposed film as is shown in Figure 5.8.

Sanger Dideoxy-DNA Sequencing

In 1977 Dr. Fred Sanger developed another commonly used method of determining the base sequence of a DNA molecule. The Sanger technique differs from the Maxam and Gilbert technique in that the former relies on interrupted DNA synthesis, while the latter relies on breakage of the DNA molecule. Both techniques, however, rely on the selective modification of specific bases and the electrophoretic separation of the resulting DNA fragments.

The Sanger technique has five basic steps:

1. Preparation of modified versions of each of the four relevant bases.

2. Preparation of a radioactively labelled DNA **primer** used to initiate DNA synthesis.

3. Synthesis of complementary DNA molecules in the presence of the modified deoxyribonucleotides.

4. Denaturation to separate the newly synthesized DNA strand from the template strand.

5. Electrophoretic separation of the resulting DNA fragments.

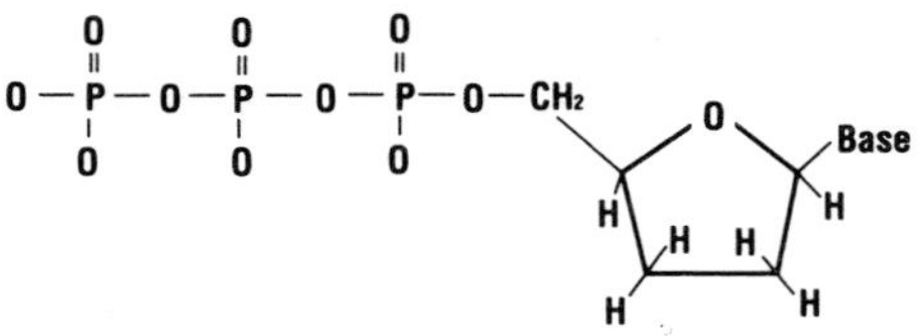

Figure 5.9
Structure of a **dideoxynucleotide** (ddNTP). The lack of a 3' hydroxyl group (-OH) prevents bond formation with adjacent deoxynucleotides. As a result, elongation of the polynucleotide chain terminates at the site of the incorporated dideoxynucleotide.

The Sanger technique relies on the fact that a growing DNA molecule requires the specific chemical configuration of a nucleotide to be in place before the next nucleotide can be added to the growing polynucleotide chain. A nucleotide, modified in some way, might not be able to bond to the next nucleotide in the molecule. This means that the process of DNA elongation (Chapter One) could be halted prematurely, producing only DNA fragments rather than a whole molecule.

The Sanger technique requires that a supply of each of the four deoxyribonucleotides be modified chemically, so that incorporation of the modified structure makes continued elongation of a DNA molecule impossible. The modified deoxyribonucleotides have structures that are slightly different from the usual deoxyribonucleotides in that they are missing the 3' hydroxyl (–OH) group. The modified bases, called **2',3'-dideoxynucleotides,** are often abbreviated as **ddNTP**. The modified forms of the individual bases are called **dideoxyadenine, dideoxycytosine, dideoxyguanine,** and **dideoxythymine** (Figure 5.9). None of these ddNTPs allow elongation of the growing DNA molecule due to the lack of the 3' –OH group, necessary for bonding between the adjacent deoxynucleotides in a DNA molecule. Thus, incorporation of a ddNTP into a growing DNA polynucleotide molecule causes a premature halt to the elongation process.

To determine the base sequence of a DNA molecule by the Sanger method (Figure 5.10), the DNA molecule is heated to break the hydrogen bonds which hold the two strands together in the DNA double helix. The resulting single strand is then bonded to a complementary primer sequence which has been radioactively labelled. This primer sequence acts to prime the synthesis of a new polynucleotide DNA strand which is complementary to the original single DNA strand. Copies of the primer-DNA complex are then divided into four aliquots. In addition to the four unmodified nucleotides and the DNA polymerase necessary for the synthesis of the new complementary strand, each aliquot receives a small amount of one of the ddNTP preparations. The concentration of the added ddNTP is carefully controlled so that these modified nucleotides will only be incorporated into some of the possible sites on the growing DNA strands. Let's assume, for example, that there are fifteen sites which, being paired with the thymine nucleotide on the template DNA strand, should contain adenine. The

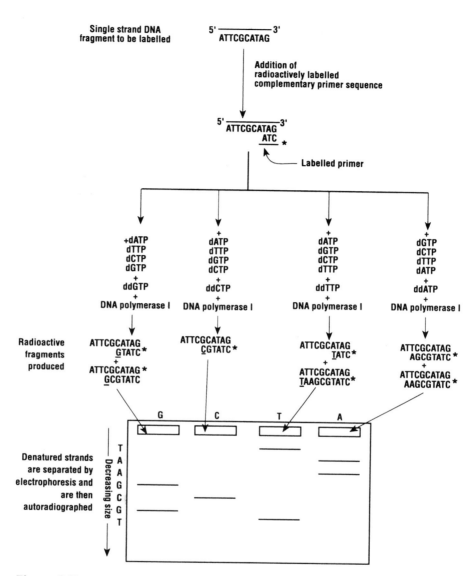

Figure 5.10
Sanger technique for DNA sequence determination. A single-stranded DNA fragment to be sequenced is allowed to bond with a small, radioactively labelled, complementary fragment called a **primer.** A preparation of primer-bound fragments is then divided into four aliquots. Each aliquot is mixed with the four nucleotides needed for synthesis of a complementary DNA strand along with one dideoxynucleotide (ddNTP), and DNA polymerase I. The resulting newly synthesized DNA strand terminates wherever a dideoxynucleotide incorporates into the sequence. Denaturation causes the newly synthesized strands to separate from the DNA strands which acted as templates for their formation. The radioactively labelled ddNTP-containing strands are length-separated by electrophoresis and subject to autoradiography. By relating the size of any individual fragment to the ddNTP which it contains, one can determine the precise position of all bases in the original DNA fragment. Thus, the base sequence of the mapped fragment can be read directly from the autoradiograph.

presence of ddATP in the reaction mixture means that this modified dideoxynucleotide will be incorporated into one of those sites. The actual site—site #1, #2, #3 or . . . #15—will be determined by chance with each site having an approximately equal probability of incorporating ddATP in place of adenine. The actual site at which a modified base will be incorporated cannot be predicted. Eventually, however, all of the possible sites will have been occupied by a modified base in a least some copies of the newly synthesized molecules.

As described above, elongation of the molecule stops wherever a ddNTP incorporates in the place of a normal nucleotide. The DNA fragments resulting from this abortive elongation process will be radioactively labelled because of the label that was attached to the primer molecule used to initiate the synthesis of the new DNA strand. The DNA fragments are then treated to cause denaturation so that the newly synthesized strand can be separated from the template DNA strand. The resulting single-stranded fragments can be size-separated by polyacrylamide gel electrophoresis, and the band pattern of the gel determined by autoradiography. Once again, as in the Maxam and Gilbert technique, the base sequence can be read directly from the exposed photographic film.

REVIEW AND SUMMARY

1. Restriction enzymes cleave lengths of DNA in a reproducible fashion. DNA molecules can therefore be identified by the pattern of fragments which are produced after cleavage by various enzymes. Identification of the recognition sequences along a DNA molecule is used to produce a **restriction map.**

2. DNA fragments, produced by the treatment of DNA molecules with restriction endonucleases, can be separated on the basis of size by the technique of **gel electrophoresis.** Agarose gel electrophoresis separates DNA fragments which differ from one another by at least 30 to 50 nucleotide pairs in length. Polyacrylamide gel electrophoresis separates fragments which differ by only a single nucleotide.

3. Gel electrophoresis functions by creating an electrostatic field which causes DNA fragments to migrate through a molecular sieve created by dissolving agarose or a similar substance in a salt water mixture and allowing the resultant liquid to solidify. DNA fragments move through the sieve with a speed that depends on both their size and electric charge, with the distance migrated being inversely proportional to the molecular size.

4. Although the base sequences of the chromosomes of different members of the same species closely resemble one another, some differences, called **polymorphisms,** exist. Non-coding polymorphisms have little, if any, effect on the functioning and overall appearance of the individual in question. Each individual has enough polymorphic sites to render his or her DNA unique. An analysis of the various polymorphisms in any one organism will provides a unique profile which can be used to identify that organism.

5. DNA polymorphisms can be identified by the ability of various restriction enzymes to cleave DNA in the vicinity of the polymorphism. The variable sizes of the resulting DNA fragments depends on the possible effect of a polymorphism on an enzyme recognition site. These variable-sized fragments are known as **Restriction Fragment Length Polymorphisms** or RFLPs.

6. RFLPs pass from generation to generation in much the same way as any other polymorphic phenotypic trait such as blood type or eye color. Therefore RFLPs can be used to trace passage of disease states among the members of a multi-generational extended family.

7. The base sequence of DNA molecules can be determined using a variety of techniques. The Maxam and Gilbert technique depends on the selective degradation of specific bases within the molecule to be sequenced. This degradation results in the production of a large population of DNA fragments which are separated using polyacrylamide gel electrophoresis. The resulting band pattern, determined by autoradiography, is used to determine the base sequence.

8. The Sanger Dideoxy-DNA sequencing technique relies on the interruption of DNA synthesis. By adding a carefully chosen amount of modified dideoxynucleotides to a reaction mixture of DNA subunits, it is possible to halt DNA chain elongation at various base sites. Since the elongation is stopped at random locations, a wide variety of DNA fragments will be produced. Polyacrylamide gel electrophoresis separates these fragments. The band pattern of the gel, determined by autoradiography, yields the base sequence of the DNA.

Cloning a gene: How to photocopy heredity

CHAPTER 6

Scientists working in a research laboratory today have a wide variety of catalogs available to them. A marvelous selection of equipment, chemicals, and other experimental agents can be purchased to facilitate scientific research. Among the common items such as laboratory centrifuges, test tubes, and containers of chemicals, one catalog offers something most interesting and unusual—some amazing mice! The chromosomes of these mice have been thoroughly analyzed. When you purchase a mouse from this catalog, you know a great deal about its hereditary background, its pedigree, and what physical characteristics it will develop. For example, you can purchase a mouse guaranteed to develop a particular genetic disease at a certain age. But perhaps most amazing, you can purchase 12 genetically identical copies of any mouse that you order. The company that produces this particular catalog learned to manipulate the breeding patterns of various mouse populations so that they can produce an almost unlimited number of mice, all of which have identical sets of chromosomes: they can produce genetic "xerox" copies of mice!

These genetically identical mice exemplify a type of population referred to as a **clone.** Clone, a complex term, not only refers to populations of cells or organisms with identical hereditary components (like the mice in the example above), but also refers to populations of identical recombinant DNA molecules. The term can also be used as a verb: to **clone** a molecule means to make multiple identical copies of that molecule, while to clone an animal population means to create genetically identical animals through a process of intensive inbreeding.

Growth of the science of biotechnology relies directly on the development of techniques that allow the cloning of biologically active molecules. In fact, many scientists today consider molecular cloning the single most important development in the growth of biotechnological research. Only through the techniques of cloning does a DNA molecule become accessible to the study and manipulation of biotechnology and genetic engineering. Let's see why.

Cloning a Gene: How to Photocopy Heredity

OVERVIEW

In most genetic engineering experiments a gene—called the **target gene**—encodes genetic information that is of interest to the scientist. For example: perhaps in the future a scientist will discover a new perspective on sickle-cell anemia, a disease resulting from faulty information contained in mutant copies of the gene that encodes one of the two types of polypeptide chains which make up the human hemoglobin molecule. To bring about a cure, this scientist might look for the information necessary to transcribe and translate normal hemoglobin molecules. With information in the form of a normal hemoglobin gene—the **target gene**—perhaps the scientist could make normal hemoglobin molecules which might then be used in some way to treat sickle-cell anemia patients, or he might even try inserting the gene directly into his patients' cells.

Another example, taken from a series of events which occurred in the early 1980s, concerns a group of scientists who were interested in diabetic patients—people who produce little or no insulin to control their blood sugar levels. These scientists wanted to isolate the gene which encodes human insulin—the target gene—and use the information contained in that gene to force bacterial cells to make a supply of the human hormone. They accomplished this goal and doctors now use this bacterially-produced human hormone to control the blood sugar levels of the majority of human diabetics!

In each of the previous examples and, in fact, when dealing with the isolation of any target gene, the scientist faces a double-ended problem: 1) Identification and purification of a single target gene can be a tedious and time consuming task. Yet, 2) to make efficient use of the genetic information in the target gene, the scientist must have many, many copies of the gene. Thus, the problem: although it can be difficult to isolate a target gene, the scientist needs multiple copies of it. Here gene cloning comes into action. If the scientist can isolate just one copy of the target gene, it can be cloned to yield an almost limitless number of copies of the original gene.

This chapter discusses the basic steps used by scientists to clone a DNA molecule: a series of DNA fragments can be isolated from the whole genome, protected from degradation by cellular enzymes, and caused to undergo multiple rounds of replication to result in a true **clonal** population of DNA molecules. This chapter also answers the following questions:

1. What can we use as a molecular photocopy machine to generate many accurate copies of our target gene?

2. What is a cosmid, and why is it important?

3. What is a gene library?

GENETIC INFORMATION CAN BE STORED IN A GENE LIBRARY

Let's imagine that you are writing a report for your history class. You've chosen the Boston Tea Party as your subject and begin to gather relevant information, probably in the library (Figure 6.1). The library, truly an amazing place, is filled with all kinds of information, only some of which relates to the Boston Tea Party.

Figure 6.1
Libraries contain information pertaining to a wide variety of subjects. Efficient use of a library requires that relevant information be separated from irrelevant information.

All of the information in the library—both related and unrelated to your subject—is divided up into discrete units called **books.** Upon entering the library you search through the books—you **screen** them—looking for your target books, those containing information about the Boston Tea Party.

Scientists also create libraries that contain information—**genetic** information. Instead of containing books, gene libraries contain DNA fragments. Just as books are protected by their covers, special molecules, called **vectors** (described later in this chapter), protect the DNA fragments in a gene library. Just as you can easily make copies of pages from a reading library, a researcher can make copies of DNA fragments—in the form of molecular clones—from a gene library. Therefore, a gene library can contain millions of copies of a target gene, even though only two copies exist in each somatic cell of the donor organism. This ability to generate large numbers of copies of individual DNA fragments makes it much easier to identify a particular target gene when it is in a library than when it is in the form of whole chromosomes as found in a normal cell. In fact, gene libraries are so useful that scientists wanting to isolate a particular gene often begin by constructing an appropriate library.

CONSTRUCTION OF A GENE LIBRARY: AN OVERVIEW

Although a variety of ways to construct and use a gene library exist, all of them employ four basic steps (Figure 6.2). This chapter describes the first two steps in some detail, while steps three and four will be described in chapter seven.

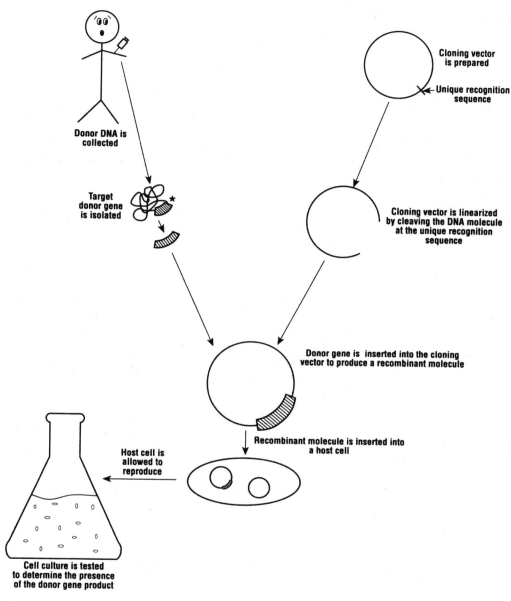

Figure 6.2
Basic steps involved in the construction of a gene library. Cells isolated from the donor are lysed and the DNA is isolated. Donor DNA is then cleaved using a restriction endonuclease, and the resulting fragments are inserted into a circular cloning vehicle which was previously linearized with the same restriction enzyme. The resulting recombinant molecule is inserted into a host cell, which permits the replication of the recombinant vector. Transformed cells are selected by the use of appropriate selective media.

1. **Selection and preparation of the donor genetic material:** The **donor genome**—all of the genetic material—from a cell likely to contain accessible genes of interest—is isolated and cut into multiple fragments using restriction endonucleases.

2. **Construction of donor DNA/cloning vector recombinant molecules:** The fragments of donor DNA are then inserted into special protective DNA molecules called cloning vectors.

3. **Insertion of the recombinant molecule into the host cell:** The donor DNA/cloning vector recombinant molecules are then inserted into appropriate **host cells,** cells with the enzymatic machinery to allow replication of the recombinant molecules.

4. **Growth of host cells and selection of transformed cells:** Having attempted to place copies of the recombinant DNA molecule into appropriate host cells, the researcher now grows those cells and performs a bit of detective work: which host cells, if any, have actually been **transformed**—that is, which host cells now contain at least one copy of the recombinant DNA molecule?

STEP ONE: SELECTION AND PREPARATION OF THE DONOR GENOME

Selection of the Donor Genome

Constructing the gene library begins with the task of choosing a suitable donor genome. We know that proteins are made as a result of the machinery of a cell following precise instructions contained in the base sequence of a gene (Chapter Two). Gene A, for example, must be present in a cell if that cell is to synthesize polypeptide A. If gene A is not present, the cell lacks the instructions necessary to produce polypeptide A. This situation parallels what would happen if you wanted to play tic-tac-toe on your computer, but accidentally loaded a blank diskette rather than one containing the game program. No program instructions on the diskette—no game! Similarly, no gene—no polypeptide product!

Let's imagine that we are physicians specializing in the treatment of diabetes. Our patients require daily injections of insulin, so cloning the gene which encodes the human insulin hormone interests us. In order to precisely locate and isolate a copy of this gene so that we can clone it, we decide to construct a library from a donor genome likely to contain our target gene.

Specialized cells in the human pancreas, called **islet cells,** produce human insulin. We therefore reason that human pancreatic islet cells must contain functional copies of the gene responsible for encoding the human insulin molecule. However, human cells contain approximately 100,000 genes in addition to the insulin gene, and the 100,000 genes represent only about five percent of the total DNA in the human genome. How can we identify and separate the target insulin gene from all of that other irrelevant DNA in the genetically crowded human

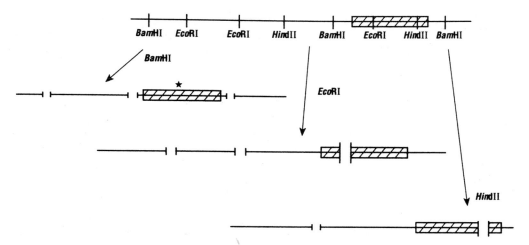

Figure 6.3
The use of multiple restriction endonucleases in the isolation of target genes increases the probability that the entire target gene will be contained, unharmed, on a single DNA fragment.

cell? Construction and use of our gene library helps us sort out this puzzle. Not only does our library allow us to magnify each individual fragment of the donor genome through the cloning process, but it enables us to break the genome into smaller, more manageable pieces through which to search. Having selected an appropriate donor genome, the next step toward construction of the library is preparation of the donor genetic material. One approach to this task deals directly with the DNA of the donor cell, while another approach works with the RNA. Let's look at these approaches one at a time.

Preparation of Donor Genetic Material: DNA

We chose a donor cell containing a gene that interests us. First, since we know the gene is located somewhere along the DNA strands that make up the genome of the donor cell, we cut samples of the entire donor genome into fragments using a series of restriction endonucleases (Figure 6.3). Each restriction enzyme cleaves the DNA at a unique set of sites and therefore produces a unique set of fragments. Each of the DNA fragments will ideally be slightly larger than an average sized gene. Since the donor genome contains the target gene, one of those DNA fragments should also contain the target gene. The use of multiple restriction endonucleases increases the probability that the target gene will be contained, unharmed and in its entirety, on a single fragment.

Preparation of Donor Genetic Material: RNA

A particular donor cell is often chosen as a source of genetic material for cloning because it synthesizes a protein of interest. Knowing about the processes of transcription and translation (Chapter Two), we deduce that mRNA molecules encoded by our target gene must also be present in such a donor cell. We can therefore use the cell's mRNA molecules as a source of genetic information for cloning a target gene. To do this we must first collect mRNA from the donor cell and then convert the information contained within the mRNA molecule back into DNA form.

ISOLATION OF mRNA FROM A DONOR CELL Whenever we isolate something specific—a special new fish for an aquarium, a child from a group of children, or a molecule from a group of molecules—we first identify something unique about our target. When we choose a new goldfish from a tank full of goldfish at the pet store, we might tell the sales person, "Please get me that one with the pointed dorsal fin and the bright reddish color." Then the sales person nets that target fish out of all of the others in the tank. Similarly, you might look for the bright green jacket your young cousin wears in order to identify her in a group of children at the playground.

In order to isolate a particular type of molecule from other molecules we must find something unique about our target. Messenger RNA molecules are unique in that most of them possess a 3' poly(A) tail consisting of 150 to 200 adenine nucleotides (Chapter Two). To isolate mRNA molecules, researchers exploit this distinguishing feature in a technique called **affinity chromatography.** This technique collects mRNA molecules via their tendency to stick to short sequences of complementary thymine deoxyribonucleotides which are attached to cellulose particles while other molecules pass by the particles. After contaminating molecules are washed away, a relatively pure population of mRNA molecules can be collected from the solid substrate.

IN DETAIL: CHROMATOGRAPHY

Chromatography is the scientific term for what happens when some water spills on a letter written with water-soluble ink—the ink runs along the paper. If the ink is made up of more than one color—for example, purple ink contains both red and blue inks—the colors often separate as they run. This happens because one color is more soluble in water than the other color. The more water-soluble color runs farther because the water carries it along more easily. The less soluble color has less tendency to run since it mixes less easily with the water.

Affinity chromatography involves three main components: a **solid substrate,** which is treated with a **chemical agent,** and a liquid which contains **materials to be separated.** Typically, the solid substrate is poured into a tube which is open on both ends and is called a **column**. The mixture which contains the target molecule along with contaminating molecules is then added to a liquid vehicle, and the resulting compound is slowly poured on top of the filled column. As the liquid filters through the solid substrate in the column, some of the molecules in the liquid have a greater tendency to stick to the substrate. Other molecules proceed more quickly through the column. In this way different components of the liquid mixture can be separated based on the length of time that it takes them to progress through the column.

By treating the solid substrate with something designed to interact strongly with mRNA molecules, we can be certain that these molecules will be hung up on the substrate while other molecules proceed through the column. Here the mRNA poly-A tail becomes useful. Adenine nucleotides bind to thymine nucleotides because they are complementary to one another. Therefore, in order to isolate mRNA molecules, the solid substrate—in this case cellulose—is treated with short linear molecules made up solely of thymine nucleotides. The resulting column is called an **oligo (dT) cellulose column.** When a collection of molecules passes over such a column, the mRNA molecules adhere to the solid substrate because of the complementary base pairing between the poly-A tails and the thymine nucleotides. The other molecules pass through the column and can be removed. Chemical treatment of the solid substrate which remains behind in the column then releases the mRNA molecules.

Target mRNA Can Be Enriched by Choice of Donor Cell

The small size of the target gene in relation to the rest of the genome challenges any attempt to clone a gene. Searching for a target gene in a donor genome compares to looking for a needle in a haystack!

Although the genomic DNA is virtually identical in most of the cells of a given organism, the relative amounts of proteins produced by those cells may vary widely from one another. For example, although pancreatic islet cells and immature blood cells have the same genetic complement, they produce vastly different predominant proteins: the islet cells predominantly produce insulin while the immature blood cells produce mostly hemoglobin.

This discrepancy can be explained by looking at the rates of transcription of various types of mRNA molecules. In the pancreatic islet cell, the insulin gene is transcribed over and over again while other genes are transcribed at much lower rates. Because of this difference in the rates of transcription, more insulin mRNA molecules are produced than any other type of mRNA, thus allowing for the increased production of insulin. Similarly, the hemoglobin gene of the immature blood cell, transcribed to a greater extent than are other genes in this cell type, results in a relative excess of both hemoglobin mRNA and hemoglobin protein. Thus, despite the similarities of the genomes, the cell types contain very different populations of mRNA molecules: certain types of mRNA molecules are **enriched** as compared with other types of mRNA molecules. Cell types which experience this type of lopsided transcription to fulfill a specialized function—such as the production of insulin or of hemoglobin—are called **differentiated cells.**

Isolation of mRNA from an appropriate differentiated cell type makes the researcher's task of target-gene identification much easier. Where two copies of a target gene might exist in the genome, 100 or more copies of the target mRNA might exist in the cytoplasm of the right differentiated cell! Nevertheless, one cannot reach into a cell and simply pick out a pure population of target mRNA molecules; the isolated population will consist of both target and non-target molecules. Despite this fact, the mRNA molecules which are isolated from the donor cell—both target and non-target—can be used as templates for the conversion of hereditary information back into DNA form.

CONVERSION OF RNA INFORMATION INTO DNA FORM

Complementary DNA—also known as **cDNA**—is a type of DNA molecule which is constructed using an RNA molecule, rather than a single strand of a DNA molecule, as a template. The synthesis of cDNA relies on the action of an enzyme called **reverse transcriptase.** This enzyme is normally found in **retroviruses**—a type of virus containing genetic material consisting of RNA in place of the more commonly found DNA. When a retrovirus infects a cell, it injects its RNA into the cytoplasm of that cell along with the reverse transcriptase enzyme. The enzyme functions to cause synthesis of a DNA molecule which is complementary to the retroviral RNA molecule. This cDNA produced from the RNA template contains virally derived genetic instructions and allows infection of the host cell to proceed.

How Is cDNA Synthesized on an mRNA Template

The mRNA molecules collected by affinity chromatography are used as templates for the synthesis of cDNA molecules. This step, in which the genetic information contained in an mRNA molecule is converted into a DNA molecule, depends on the enzymatic action of reverse transcriptase. Although a variety of specific methods are used to synthesize cDNA, only one representative method will be described here (Figure 6.4). To prepare cDNA, mRNA molecules are mixed with short **primers,** or chains of thymine deoxyribonucleotides, which bind, or **hybridize,** to the complementary 3′ poly(A) tail of the purified mRNA. The primers provide a starting place—an **initiative site**—for the action of reverse transcriptase which, in turn, causes the synthesis of a strand of DNA complementary to the mRNA template molecule.

Interestingly, once the enzyme travels the length of the mRNA template, it turns around and begins to copy the cDNA strand which it has just synthesized. It is unknown whether this event occurs in nature or whether it is an artefact caused by laboratory conditions. Regardless of the cause, however, it results in production of a single-stranded cDNA molecule which ends in a U-turn, sometimes called a **hairpin loop,** and a short double stranded sequence at the hairpin-loop end of the molecule. Chemical treatment of the mRNA/cDNA hybrid molecule then causes degradation of the template mRNA molecule. The enzyme DNA polymerase can then use the short double-stranded sequence and the free end of the hairpin loop as a primer to begin synthesis of a second DNA strand, this one complementary to the first cDNA strand. Once the double-stranded molecule is complete, an enzyme called **S1 nuclease** is added to the preparation. This enzyme, which cleaves only single-stranded molecules, cleaves the hairpin loop. This whole process results in production of a molecule which is equivalent to the gene which originally encoded the mRNA template molecule, except that it is missing any exons which would normally have been found in eukaryotic genomic DNA. Although both DNA and RNA can successfully be used as donor genetic material, techniques which rely on the use of RNA have some advantages over the use of genomic DNA itself.

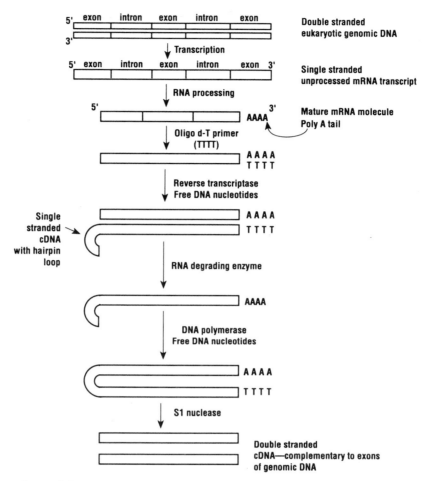

Figure 6.4
One method of producing a cDNA molecule. A mature eukaryotic mRNA molecule, consisting only of exon regions, is treated with an oligo-d(T) primer which binds to the complementary 3' poly(A) tail. The addition of reverse transcriptase and free deoxyribonucleotides results in the synthesis of a single-stranded DNA molecule, complementary to the mRNA and ending in a hairpin loop. An enzyme is added to degrade the mRNA template. Next, the addition of DNA polymerase and free DNA nucleotides result in the synthesis of a second DNA strand which is complementary to the first strand. S1 nuclease, an enzyme that selectively cleaves single-stranded DNA regions, is used to degrade the hairpin loop. The resultant double-stranded cDNA molecule is identical in sequence to the exon regions of the original genomic DNA.

cDNA Reduces the Effective Size of Eukaryotic Genes

Most eukaryotic genes contain non-coding regions in the form of introns along with the coding regions which are known as exons (Chapter Two). This arrangement of introns and exons makes many eukaryotic genes quite large, often too large to fit easily on the DNA fragments generated by restriction endonuclease cleavage.

As discussed in chapter two, both the introns and the exons of a eukaryotic gene are transcribed into RNA. However, during the process of post-transcriptional modification, the non-coding intron regions are removed from the immature RNA molecule, and the remaining exons are spliced together to produce a smaller mature mRNA molecule (Figure 2.8). This mature mRNA molecule is then transported out of the nucleus for the process of translation.

Use of the mature mRNA molecule—composed of exons alone—as a template for the production of cDNA results in production of a DNA fragment that is smaller and more easily handled than the original genomic DNA, which consists of both introns and exons.

Use of cDNA Avoids the Need for Intron Splicing

Once gene fragments which may contain the target gene have been isolated—either from genomic DNA or from cDNA—they must be protected from degradation and inserted into an appropriate host cell which will enable replication of the cloning vector and the target gene. Some of the most often used host cell types lack the enzymes needed for removal of introns and splicing of the remaining exons. Complementary cDNA, often useful as a donor material, alleviates the need for these processes (Figure 6.5).

STEP TWO: DNA FRAGMENTS ARE PLACED INTO CLONING VECTORS

In order to identify the genetic information contained on any given DNA fragment, whether it was isolated from genomic DNA or from cDNA molecules, the fragment must be inserted into a host cell that has the ability to replicate the cloning vector containing the DNA fragment. Just as you rely on vehicles—bicycles, airplanes, or automobiles—to transport you any great distance, DNA fragments must also rely on vehicles. **Cloning vectors,** the DNA molecules used as vehicles to carry DNA fragments, serve to protect the fragment from degradation and act as a container as it is transported to its final destination: the interior of a carefully chosen host cell.

Restriction Enzymes Permit Recombination of Fragment and Vector

The recombination of a DNA fragment and a cloning vector relies on the action of restriction endonucleases. You will recall from our discussion of these special enzymes that they cleave a DNA molecule only in the vicinity of particular recognition sequences of bases. In addition, enzymatic cleavage often results in the formation of sticky ends that facilitate the molecular bonding that can occur between DNA fragments from two or more different sources (Chapter Four). By cleaving both the cloning vector and the DNA fragment which is to be inserted into the cloning vector with the same restriction enzyme, sticky ends can be generated on the two DNA molecules. Bonding which occurs between these sticky ends results in the joining of the DNA fragment and the vector molecule. The resulting product is a vector molecule that contains a DNA insert (Figure 6.6).

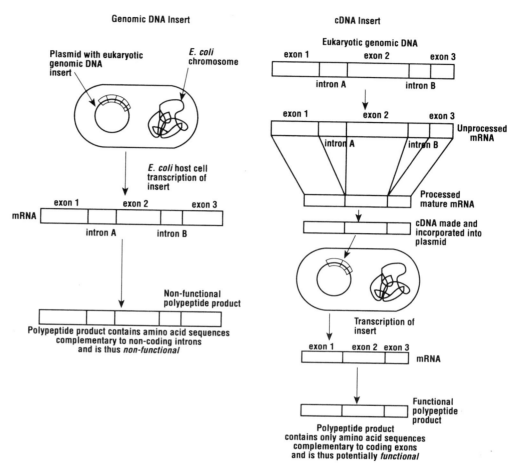

Figure 6.5
Use of eukaryotic genomic DNA *vs* cDNA as donor DNA in cloning protocols. If genomic DNA consisting of both introns and exons is placed in a cloning vector and inserted into a host cell capable of expressing the insert, the resulting polypeptide product will be complementary to both coding and non-coding regions of the cloned gene. Such a product is likely to be non-functional. In contrast, if cDNA, which consists only of exons, is inserted into a host cell capable of expressing the insert, the resulting polypeptide product will be complementary to exons alone. Such a product would be potentially functional.

Cloning Vectors Have Special Characteristics

The term "cloning vector" refers to any molecule, usually circular, which protects a DNA fragment, inserts it into a host cell, and facilitates replication of the DNA fragment once it is inside the host cell. To be most useful, a cloning vector should have at least one, and ideally more, restriction enzyme recognition sites which occur only once along the entire cloning vector molecule. In other words, ideally there should be only one *Bam*HI site or only one *Eco*RI site located on the cloning vector. Such restriction sites, called **unique** sites, allow the scientist,

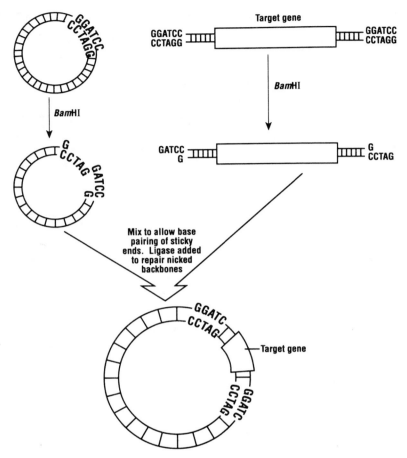

Figure 6.6
Using the same restriction enzyme, both to isolate the target gene which is to be cloned and to linearize the cloning vector, always generates DNA fragments with complementary sticky ends. Mixing the two types of fragments with one another results in the formation of recombinant molecules consisting of two or more fragments joined by their sticky ends. Some of those recombinant molecules consist of the target gene inserted into the cleavage site of a linearized cloning vector. This type of recombinant molecule is often used in cloning protocols.

armed with restriction endonucleases, to insert DNA fragments into the circular cloning vector without destroying the integrity of the vector by breaking it into multiple pieces. If a specific restriction site exists in a particular vector more than once, cleavage of the vector results in its destruction because vector fragments would separate from one another. In contrast, a unique site means that cleavage will simply convert a circular vector molecule into a linear one. No fragment of the vector separates from the remainder of the vector (Figure 6.7).

In addition, an efficient and useful cloning vector should have at least one readily detectable property—known as a **selectable marker**—to confer upon the host cell once the vector has been inserted into the host cell. Presence of

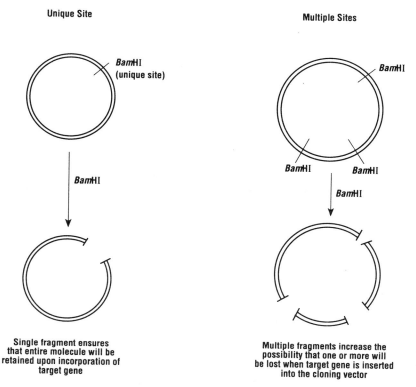

Figure 6.7
Multiple restriction sites on a circular cloning vector may lead to the loss of one or more fragments from the vector upon enzymatic cleavage.

the selectable marker—tetracycline resistance, for example—can then indicate the presence of the plasmid itself within the host cell.

Two basic types of molecules fulfill these requirements and thus are useful as cloning vectors. One molecule is called a plasmid and the other is the chromosome of a bacteriophage. Let's look at these cloning vectors one at a time.

THE PLASMID AS A CLONING VECTOR

Plasmids, double-stranded circular DNA molecules, are most often found in bacterial cells along with, but separate from, the circular bacterial chromosome. Important in their role as a cloning vectors, plasmids accept the insertion of up to approximately 5,000 base pairs of foreign DNA and still retain the structure and functions associated with the original plasmid. Since a plasmid is not physically linked to the bacterial chromosome, it can sometimes be lost from the cell during cell division. In nature, individual bacterial cells have varying numbers of plasmids, ranging from one to several hundred per cell. Plasmids vary in size from a relatively small 1,000 base pairs to more than 300,000 base pairs in length. In contrast, the single circular chromosome of an *E. coli* cell is composed of approximately 5 million base pairs and the 46 linear chromosomes of a human cell total 3 billion base pairs.

Plasmids Can Replicate

Replication of a plasmid molecule requires that it contains an **origin of replication**—a site at which the replication process is able to begin. Plasmid replication can be either dependent upon or independent of replication of the bacterial host cell chromosome. A plasmid whose replication depends on replication of the bacterial chromosome—a **stringent plasmid**—generally exists as a single copy within the cell. Conversely, a plasmid whose replication is independent of host cell replication is called a **relaxed plasmid** and exists at levels ranging from approximately ten to four hundred copies per cell.

Plasmids Carry Genetic Information

Plasmids carry a variety of genes, often including the genes which govern their own transmission, from cell to cell during bacterial mating or **conjugation.** In this process two organisms, usually single celled, temporarily come together and exchange genetic material, including plasmid molecules. The ability of a plasmid to be transferred among cells is helpful, although not required, in the cloning process because it results in the acquisition of the recombinant plasmid by additional host cells. This, in turn, allows an increased level of plasmid replication as the host cells themselves divide.

In addition, as described in chapter four, plasmids often carry genes for selectable markers, including some that confer specific antibiotic resistance on the cells that carry them. For example, some plasmids carry a gene that encodes a protein which enables a cell to resist the normally lethal effect of the antibiotic known as tetracycline. A cell without this plasmid gene would die in the presence of tetracycline. If a plasmid containing the tetracycline resistance gene is inserted into the cell, it enables that cell to survive despite the presence of that particular antibiotic. Similarly, other plasmids carry other antibiotic-resistance genes.

An Example of a Plasmid Cloning Vector: pBR322

One of the first commonly used plasmids in genetic engineering is called **pBR322**. This plasmid, engineered from an *E. coli*-derived plasmid, consists of 4,362 base pairs and has been completely sequenced. pBR322 carries genes which confer both tetracycline and ampicillin antibiotic resistance on any cell which contains the plasmid. In addition, multiple unique restriction sites on the pBR322 plasmid include several sites which are located within the two drug-resistance genes. These especially useful sites allow the detection, by a phenomenon known as **insertional inactivation** (Figure 6.8), of any DNA fragment which is inserted into the plasmid at these sites.

Insertional inactivation means that if a gene fragment is inserted into one of the unique restriction sites in a gene for drug resistance, it can be detected because it will cause the drug-resistance gene to become nonfunctional. In other words, if a DNA fragment is inserted into the *Eco*RI site which lies within the pBR322 gene for tetracycline resistance, that gene no longer functions although the gene for ampicillin resistance remains functional. A clonal population of

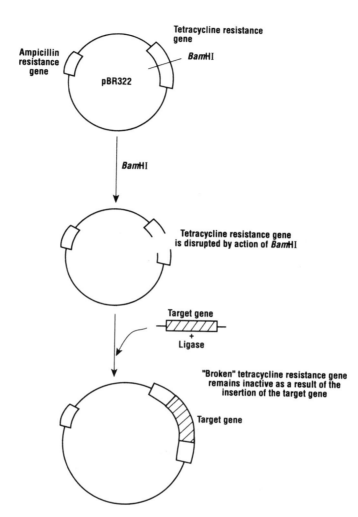

Figure 6.8
Insertional inactivation occurs when a DNA fragment is inserted into a restriction site located within a potentially active gene, such as that for tetracycline resistance. Insertion can be detected by loss of the function encoded by the interrupted gene.

cells can be divided with part of the population being grown in the presence of tetracycline, and another part of the population being grown in the presence of ampicillin. A population of cells which has undergone insertional inactivation of the tetracycline gene will be unable to grow in tetracycline-supplemented medium, but will grow in ampicillin-supplemented medium. This growth pattern indicates insertion of a DNA fragment within the tetracycline gene and therefore indicates successful insertion of a DNA fragment into the cloning vehicle.

About 50 copies of pBR322 plasmids normally exist in each healthy bacterial host cell. The number of copies can be markedly increased, however, by a process known as **amplification**. In this process, a cell containing a relaxed plasmid is treated with a drug to inhibit protein synthesis. As a result of this treatment, the cell stops replicating. The relaxed pBR322 plasmid, however, continues to replicate despite the drug treatment since the replication of relaxed plasmids is independent of cell replication and does not require protein synthesis. For example, addition of the protein-synthesis inhibitor **chloramphenicol** causes the

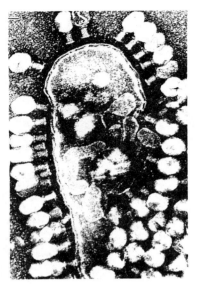

A

number of pBR322 plasmid molecules to increase to as many as 3,000 per cell! This amplification process results in an increased ratio of plasmid DNA to chromosomal DNA, thus making it easier to isolate the plasmid DNA. Amplification, extremely beneficial in that it results in an increase in the amount of plasmid DNA, also increases the amount of donor insert DNA. If one wants to make lots of copies of the insert DNA by cloning, amplification processes become a useful tool!

THE BACTERIOPHAGE AS A CLONING VECTOR

A **bacteriophage,** or more commonly a **phage,** is a virus particle that infects a bacterial cell. A phage particle usually consists of a **phage head** which contains the genetic material and a **tail** through which the genetic material is injected into the host cell. One of the most widely studied of these phage particles, called **bacteriophage Lambda,** has become a commonly used cloning vector. Approximately 49,000 base pairs compose the Lambda phage chromosome, a double-stranded linear DNA molecule. The phage chromosome contains about 60 functional genes. The 5' end of each of the two strands of the DNA molecule terminates in a single stranded segments of about 12 nucleotides. These single stranded segments, called **cos sites,** complement one another. Because of their complementarity the *cos* sites can bind together to circularize the DNA molecule.

Lambda phage generally begins its infection of a bacterial cell by binding to a receptor protein on the bacterial cell wall (Figure 6.9a). The linear phage DNA is injected into the host bacterial cell where it becomes circularized via the binding of its *cos* sites. At this point the phage DNA may take one of two routes (Figure 6.9b).

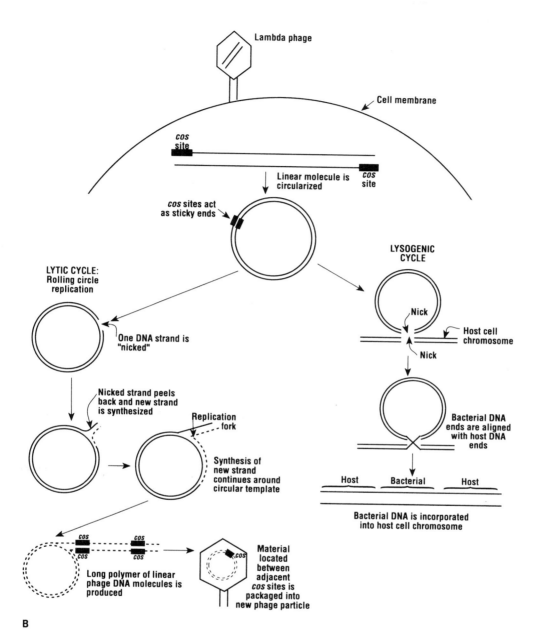

Figure 6.9a and b
Bacteriophage lambda contains a single double-stranded DNA molecule with two 5' single-stranded sticky ends, known as *cos* sites, which allow circularization of the molecule.
(a) Phage attaches to the membrane of a host cell and injects its DNA molecule, in linear form, into the cell's interior. (b) Once inside, the linear molecule circularizes via the *cos* sites. The circularized molecule can then undergo either the lytic cycle or the lysogenic cycle of infection. In the lytic cycle of infection, the phage DNA molecule reproduces many times, perhaps involving the rolling-circle mode of replication. This process results in the production of a long polymer of linear phage DNA molecules, copied from the circular template, and joined to one another by their *cos* sites. The polymer cleaves at the *cos* sites and material located between two adjacent *cos* sites is packaged into a new infective phage particle. Production of multiple new phage particles eventually leads to lysis of the host cell. In the lysogenic cycle of infection, the phage DNA molecule becomes incorporated into the host chromosome and remains relatively inactive for varying amounts of time. Eventually the phage DNA activates, leaves the host chromosome, and begins the lytic cycle of infection. (From Oliver and Ward, *Dictionary of Genetic Engineering.* Copyright © 1985 Cambridge University Press, New York. Reprinted by permission.)

Bacteriophage Lambda: The Lytic Cycle

In the first route—known as the **lytic cycle**—replication of the circularized DNA occurs immediately and results in the production of multiple new phage chromosomes which are then packaged in phage heads to form new phage particles. This replication takes place in two ways. The first is the more common DNA replication process discussed in chapter one. The second way—called the **rolling circle mode**—is somewhat more unusual. In this mode, a nick is made in one strand of the circularized phage DNA molecule. The nicked end then peels back to reveal the un-nicked template strand. As replication occurs along the template strand, the nicked end peels back farther and farther. Eventually, replication proceeds around the entire circular template strand. Replication does not stop at this point, however. Instead, as replication continues to circle repeatedly around the template, a long molecule is formed. This molecule—known as a **concatemer**—consists of multiple copies of the phage DNA molecule linked end to end. Eventually the multiple copies, cut apart from one another at the *cos* sites, yield many individual linear copies of the original phage DNA molecule, each of which then circularizes via the *cos* sites. Various enzymes, derived from the genes of both phage and bacterial cells, work to package the new phage DNA molecules within phage heads, thus forming new, potentially infective phage particles. Once enclosed within the phage head, the circular DNA molecule linearizes once again. At this point the bacterial cell, which now contains many copies of the original phage particle, breaks open—**lyses**—and releases the newly formed phage particles into the surrounding environment.

Bacteriophage Lambda: The Lysogenic Cycle

In the second route—known as the **lysogenic cycle**—the phage DNA does not proceed immediately from injection into the host cell to replication and lysis of the host cell. Instead, the circularized phage DNA breaks at a specific site on the molecule. At the same time a break occurs in the bacterial chromosome. Although cellular enzymes repair both breaks, they do not repair them in the fashion one might expect. Rather than gluing the two phage DNA ends together and gluing the two bacterial DNA ends together, the enzymes join each phage strand end to one bacterial strand end. The result: the linear phage DNA molecule incorporates into the bacterial chromosome of the host cell.

The incorporated phage—called a **prophage**—is fairly inert within the host bacterial genome, and thus growth of the bacterial cell continues quite normally. As the host cell and its contents replicate and divide, the prophage replicates and divides as well. In this way, all of the descendants of a prophage-infected cell also contain a hidden prophage. Such cells function virtually normally, sometimes for many thousands of generations of cell divisions, until some event or signal causes the prophage to leave the bacterial chromosome and regain its separate circular formation. Once this occurs, the newly released phage DNA molecule begins the lytic cycle which eventually leads to lysis of the host cell.

Donor DNA Is Inserted into the Phage Chromosome

Fragments of DNA—either genomic or cDNA in origin—can be inserted into the bacteriophage chromosome in much the same way as DNA fragments can be inserted into plasmids. As long as the resulting recombinant DNA molecule has a size of approximately 45,000 to 50,000 base pairs, it can be packaged within bacteriophage heads. This results in a new phage particle containing the inserted DNA fragment as part of its chromosome: the phage particle acts as a Trojan Horse in which hides a piece of foreign DNA. Still resembling a Trojan Horse, the new phage particle can infect an appropriate host cell much as a normal phage particle would. The inserted DNA fragment, carried along with the phage chromosome, therefore enters the host cell.

RELATIVE MERITS OF PHAGE AND PLASMIDS AS CLONING VECTORS

Plasmid vectors are generally much smaller than phage vectors, so phage vectors more easily accept the insertion of relatively large DNA fragments. While plasmid vectors can accept an average insert of about 4,000 base pairs, phage vectors can accept inserts of about 20,000 base pairs. In general, however, plasmids are somewhat easier to use as cloning vectors than are phage DNA molecules. Plasmid DNA often contains unique restriction sites. In fact, scientists can construct artificial plasmids that are designed to contain particularly suitable unique restriction sites along with any other property deemed useful by the scientist. In addition, since plasmid molecules are relatively small, scientists sometimes detect the insertion of a DNA fragment simply by looking for an increase in the plasmid size. This does not work well with phage DNA because the inserted fragment represents a much smaller percentage of the total DNA in the phage as compared with the plasmid.

COSMID VECTORS: HYBRID VEHICLES

Cloning done using either plasmids or phage DNA as vectors is limited to cloning lengths of DNA that are between about 4,000 and 20,000 base pairs long. While some genes fall within this size limitation, many genes—especially eukaryotic—do not. Further, if one wants to clone a group of genes located adjacent to one another on a donor chromosome, the size quickly outstrips what is possible in either plasmid or phage cloning vectors.

Cosmids, mixed or **hybrid** vectors developed in the late 1970s to alleviate this problem, contain both phage and plasmid DNA. More specifically, cosmids contain lambda phage *cos* sites incorporated into plasmid DNA sequences. The plasmid sequences generally contain an origin of replication along with a few unique restriction sites and one or two genes which encode selectable markers. Cosmid vectors uniquely and usefully accept the insertion of between 35,000 and 45,000 base pairs of foreign DNA—significantly more than the 4,000 or so base pairs accepted by plasmid vectors or the 20,000 accepted by phage vectors!

Cloning with a cosmid vector takes advantage of the fact that DNA sequences located between two *cos* sites can be packaged into phage particles as long as the overall size of the recombinant vector is approximately 45,000 to 50,000 base pairs in length. The first step in cosmid cloning cleaves the cosmid at a site separate from the *cos* site. This kind of cleavage results in a linear cosmid molecule containing a single uncut *cos* site (Figure 6.10). Large fragments of donor DNA,

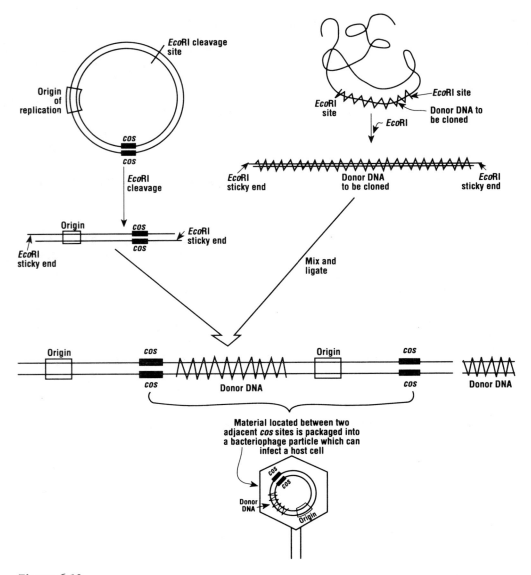

Figure 6.10
Cosmid vectors, useful plasmids which contain the lambda phage *cos* site along with one or more selectable markers, allow cloning of relatively large fragments of DNA. Cosmid cloning involves cleavage of the vector at a site separate from the *cos* site. The linearized cosmid vector then mixes with DNA fragments which are to be cloned. After ligation, some of the recombinant molecules consist of linear cosmids alternating with donor DNA fragments. This polymer simulates the polymer produced in the rolling circle mode of phage lambda replication and, as in that process material located between two adjacent *cos* sites will be packaged into new infective phage particles, used to infect appropriate host cells for cloning.

generated by restriction endonuclease activity, can then be mixed with the linear cosmid molecules. If the same enzyme cleaves both the cosmid and the donor DNA, then this mixing of fragments results in the production of complex ligated molecules which form due to the complementarity of the staggered ends at the cleavage sites. Some of these ligation products consist of a length of donor DNA flanked by two linear cosmid molecules. These ligation products closely resemble the concatemer produced during the rolling circle mode of replication. They can therefore be packaged into bacteriophage particles which can then transfer the donor DNA to appropriate host cells by the process of infection.

Now, about halfway through the construction of a gene library, we have learned about selection and preparation of donor DNA. We have seen that donor DNA can be either in the form of genomic sequences or in the form of cDNA, and that each possibility has both advantages and disadvantages. We know that donor DNA must be inserted into one of the many types of protective cloning vectors. Chapter Seven discusses the remaining steps in the construction of a gene library.

REVIEW AND SUMMARY

1. **Gene libraries** store genetic information. These libraries are composed of fragments of donor DNA which are protected by insertion into cloning vectors. Taken together, all of the inserts in the recombinant molecules of a library represent the entire genome of the donor organism. Recombinant cloning vectors from a library, inserted into host cells, allow replication of the donor genetic material.

2. Donor genetic information comes primarily from two sources: **genomic DNA or cDNA.** Genomic DNA is the entire complement of DNA in a donor cell. Eukaryotic genes composed of genomic DNA consist of noncoding introns interspersed among coding exons. Genes composed of cDNA are created using reverse transcriptase to synthesize a DNA copy of an mRNA template. Since the mRNA template has already undergone post-transcriptional modification, it consists solely of coding exons. Therefore, cDNA differs from genomic DNA in that it is composed of exons alone.

3. Genomic DNA and cDNA each have particular advantages as donor genetic material. Genomic DNA represents all of the genetic material contained within a given donor cell. Conversely, cDNA represents only those genes which were being transcribed at the time of mRNA collection. In addition, use of cDNA reduces the effective size of a eukaryotic gene since it represents only the exon portion of that gene. The lack of introns in eukaryotic cDNA allows its successful translation by bacterial cells, as these cells are unable to remove non-coding introns from the mRNA which is transcribed from a eukaryotic genomic DNA fragment.

4. Use of cDNA can make it easier to identify a particular target gene than does use of genomic DNA. Although the genomic DNA in all cells of a given organism is basically identical, the mRNA differs widely in cells with different functions. By isolating mRNA from appropriate cells—for example, the isolation of mRNA complementary to the insulin gene from pancreatic islet cells—the probability of locating a target molecule increases.

5. Messenger RNA is isolated from a host cell by passing a preparation of nucleic acids over a cellulose column which has been treated with short lengths of thymine deoxyribonucleotides. This type of column is called an **oligo-dT cellulose column.** The poly(A) tail which characterizes the mRNA molecules binds to the complementary thymine nucleotides attached to the cellulose of the column. In this way, the mRNA molecules remain within the column while other contaminating nucleic acids pass through the column. The bound mRNA molecules can then be chemically removed from the cellulose.

6. Complementary DNA, synthesized from mRNA templates using the retroviral enzyme called reverse transcriptase, differs from eukaryotic genomic DNA in that it is composed solely of coding exon regions. Conversely, genomic DNA consists of non-coding introns interspersed among the coding exon regions. Therefore, cDNA gene sequences are smaller than the equivalent genomic gene sequences.

7. Donor genetic sequences, either genomic or cDNA in origin, must be protected from degradation and ferried into appropriate host cells by **cloning vectors.** Cloning vectors are lengths of DNA which generally carry unique recognition sequences, selectable markers, and can replicate. The two primary types of cloning vectors are **bacterial plasmids** and **bacteriophage chromosomes.** A third type of vector, called a **cosmid,** consists of both phage and plasmid DNA.

8. Plasmid vectors, generally significantly smaller than the circular bacterial chromosome, often contain genes for antibiotic resistance, and genes which govern their transmission from cell to cell during the process of conjugation. The replication of plasmids can be either tied to replication of the host chromosome—**stringent plasmid**—or independent of host chromosome replication—**relaxed plasmids.** The number of relaxed plasmids per host cell increases by a technique known as **amplification.**

9. Bacteriophage vectors have linear double-stranded DNA molecules which are flanked by complementary single-stranded sequences of bases known as *cos* **sites.** The *cos* sites can bind to one another, thus circularizing the phage chromosome. Phage particles inject their DNA into host bacterial cells in the process of infection. The phage DNA either immediately directs the synthesis of new phage particles in the **lytic process** or becomes relatively inert through incorporation into the host chromosome in the **lysogenic process.**

10. Donor DNA fragments of approximately 4,000 and 20,000 base pairs or fewer can be incorporated into the plasmid or phage vector, respectively, with the use of restriction endonucleases. Once incorporated, the foreign DNA, carried along with the vector DNA, also replicates along with the vector DNA.

11. Cosmid vectors represent hybrid vectors consisting of the phage *cos* site incorporated into a plasmid molecule. Cosmid vectors can accept donor DNA fragments, cleaved with an endonuclease, between 35,000 and 45,000 base pairs in length. The same endonuclease then linearizes the cosmid in such a way as to leave the *cos* site intact. Both the cosmid and the donor fragments are mixed and allowed to ligate. Some ligation products consist of a donor fragment flanked by two cosmids and therefore by two *cos* sites. Material located between two *cos* sites can be packaged into a phage particle. Therefore, the donor DNA can be packaged into a phage particle ready to infect a host cell as a normal phage particle might. The donor DNA can thus be inserted into a host cell via the process of infection.

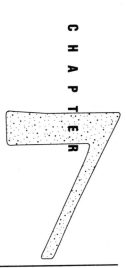

Cloning–part two: Using a gene library

Computers affect almost all aspects of modern daily life. Our driver's licenses are listed on computers along with our personal driving histories. Computers help cashiers tally our purchases at the grocery store and even interrupt our dinners to call us on the phone and tell us about great new deals in life insurance. Computers—with an enormous variety of capabilities—are everywhere.

Although they seem extremely intelligent, in fact computers are only collections of hardware that follow sets of instructions. Those instructions—called **software programs**—must be fed into the computer in order for that computer to do something as simple as calculate the sum of two plus two (Figure 7.1). If you want your computer to translate a list of words from English into Spanish, you must first tell the computer how to accomplish that task by inserting a diskette containing the appropriate instructions. Similarly, if you want your computer to help with your calculus homework, you first instruct the computer in the rules and functions of calculus by inserting a diskette with those instructions. By providing your computer with the appropriate software information, you cause the hardware of your computer to perform a great many tasks.

Molecular cloning is much the same: the computer represents the appropriate host cell while the information contained on the inserted diskettes represents the genetic material contained within the cloning vector. Just as the computer follows instructions on the inserted diskette to produce a final product—your calculus homework, for example—the cell follows instructions on the inserted cloning vector to produce a final product—molecules of human insulin, for example.

OVERVIEW We saw that many cloning experiments involve the production of gene libraries, all constructed using four basic steps (Chapter Six): 1) selection and preparation of donor genetic material, 2) construction of donor DNA/cloning vector recombinant molecules, 3) insertion of the recombinant molecule into the host cell and 4) growth of host cells and selection of transformed cells.

Figure 7.1
Computers, like cells, carry out specific functions if they have access to the appropriate sets of instructions in the form of software programs.

Chapter Six discussed steps one and two in some detail by describing the isolation of donor DNA fragments and their insertion into cloning vectors. Chapter Seven describes what happens to that donor DNA/cloning vector recombinant molecule, how the recombinant vector is inserted into an appropriate host cell, and the examination of potential host cells to determine which ones have actually incorporated the recombinant vector molecule. In addition, it answers the following questions:

1. How can something as small as a DNA molecule be inserted into a host cell?

2. What does the term "shotgun" have to do with gene cloning?

3. How can an antibiotic identify a host cell which has incorporated a recombinant vector molecule?

4. How can we find a single, specific, target gene and separate it from all of the other genes in a genome?

DIFFERENT KINDS OF GENE LIBRARIES

There Are Two Sources of DNA for Gene Libraries

As we discussed in Chapter Six, two primary sources of donor genetic material can be used to construct a gene library: the genomic DNA of almost any cell and the mRNA/cDNA of a differentiated cell. Regardless of which source of donor

DNA the scientist chooses, fragments of DNA will be inserted into carefully chosen cloning vectors. The collection of cloning vectors which contains inserts representing an entire set of donor genetic material composes a gene library.

Genomic Libraries Represent the Entire Donor Genome

A library constructed using fragments of genomic DNA represents the entire genetic complement, both transcribed and untranscribed, of the donor cell. Genomic libraries, often constructed using phage cloning vectors, grow as clear or semi-clear **plaques** or holes on a layer of host bacterial cells. Each plaque develops because of host cell lysis due to the phage infection. Plaques grow in size as phage particles released during lysis go on to infect nearby cells, which also eventually lyse, thus beginning the chain reaction which leads to plaque growth. Phage vectors are often chosen for gene library construction because they are capable of accommodating relatively large donor DNA inserts. Therefore, fewer insert/vector recombinant molecules are required to include a combination of fragments which represents the entire donor genome in the cloning procedure. In one technique, often used in the construction of a genomic library and referred to as **shotgun cloning,** the entire donor genome is enzymatically cut into thousands of random fragments, each of which is inserted into a cloning vector. The term "shotgun" may refer to the fact that a shotgun may fire many shots before it hits its target. Similarly, shotgun cloning typically involves the production and examination of many DNA fragments in order to locate one particular target gene.

cDNA Libraries Represent Only Transcribed Genes

A genomic library obviously differs from a cDNA library in that it represents a complete genome consisting of both transcribed and non-transcribed donor DNA sequences, while a cDNA library represents only that subset of genes being transcribed when the mRNA was collected for use in library production.

Since a cDNA library can be constructed from the mRNA of a differentiated cell type, certain target gene sequences can be enriched in this type of library as compared with a genomic library. Because a cDNA library does not represent the entire donor genome but instead is derived from mRNA populations which may have been selectively enriched or amplified through differentiation, it should be easier to identify a specific target using this type of library—but there is no such thing as a free ride. No matter how carefully the mRNA populations are isolated, non-target RNA molecules always contaminate our cDNA library to some extent.

Some Libraries Result in Expression of the Inserted Gene

Some libraries—known as **expression libraries**—are constructed using a cloning vector designed to allow transcription and translation of its genetic insert by the host cell. Such a vector, called an **expression vector,** contains the minimum genetic information required to permit the expression of inserted genes,

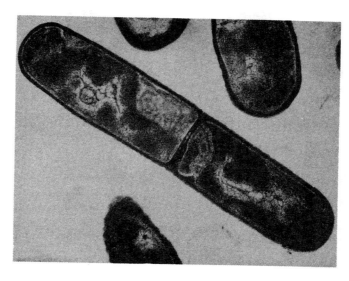

Figure 7.2
Competent cells are able to take up purified DNA molecules. Rod shaped bacteria known as *Bacilli* are examples of naturally competent cells.

as discussed in some detail in chapter eight. Expression libraries are unique in that they permit synthesis of polypeptide products encoded by the inserted foreign genes.

HOST CELLS AND THEIR TRANSFORMATION

Recombinant Vectors Must Be Inserted into Host Cells

Chapter Six described how to create a sort of genetic passenger vehicle in which a cloning vector plays the role of the automobile and a donor DNA fragment plays the role of the passenger. An automobile transports you to work or to school where you get busy and carry out your assigned function—to earn a paycheck or to study and learn. Like an automobile, a cloning vector carries its passenger to a particular location suitable for a certain function. In this case the donor DNA passenger acts as a template for the production of more DNA molecules. In order for this function to be completed, the cloning vector and its DNA passenger must actually enter a host cell. How is this accomplished?

Some Cells Are Competent Cells

Cells able to accept or take up purified DNA molecules are referred to as **competent cells.** Some cell types, naturally competent under specific nutritional conditions, include the species ***Bacillus***—a genus of rod-shaped bacteria (Figure 7.2). Other organisms, not naturally competent, must be chemically treated in order to induce a competent state. Examples of such organisms are ***E. coli*** bacteria (Figure 7.3) and **yeast.**

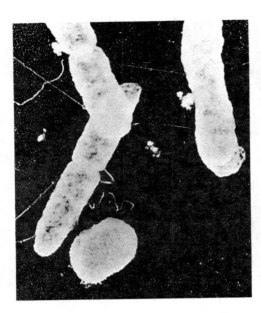

Figure 7.3
The bacterium *Escherichia coli* is an example of a cell type which is not naturally competent. This photomicrograph shows examples of rod shaped **E. coli.** Two of the cells are undergoing the reproductive process of conjugation.

E. Coli: A Versatile and Useful Host Cell

Perhaps the most commonly used host cell in the techniques of genetic engineering today, the bacterium *E. coli,* normally found in the healthy human intestine, absorbs necessary nutrients from digested materials in the gut. *E. coli* cells reproduce both sexually and asexually. The asexual process of **binary fission** occurs when the bacterial genetic material, both chromosomal and plasmid in origin, replicates. The parent cell then divides equally into two parts, with each part incorporating one set of genetic material. Thus, a single parent cell produces two identical daughter cells. In **sexual reproduction,** cells of two different bacterial mating types join one another by the formation of a fine hair-like structure called a **pilus.** Scientists think that a cytoplasmic bridge then forms through which genetic material from each cell is exchanged. After completion of the genetic exchange, the cells—each with a modified genetic complement—separate.

As a prokaryote, the 5 million base pairs which make up the *E. coli* bacterial chromosome are not found within a distinct nuclear membrane. However, thorough studies have identified both the locations and products of many of the genes on the *E. coli* chromosome. Such detailed knowledge of the *E. coli* genome and its polypeptide products certainly contributed toward making *E. coli* one of the most common host cell types in current research techniques.

Growth of E. Coli Cells

A number of well characterized populations—known as **strains**—of *E. coli* are used in modern laboratory research. Although some differences exist from strain to strain, they retain some important similarities. The majority of strains commonly used in laboratory research can be grown in a defined mix of components known as **minimal medium.** Minimal medium consists of some inorganic compounds, trace amounts of several ions, and one or more carbon-containing organic compounds which serve to provide the bacterial cells with both the carbon atoms and the energy they need to grow and reproduce.

E. coli cells growing at 37°C—normal human body temperature—in a fresh supply of minimal medium normally double in number approximately every sixty minutes. The growth of *E. coli* bacterial cell populations—known as **cultures**—speeds up when a variety of materials are added to minimal medium so that the growing cells do not expend energy to synthesize those materials. These components include amino acids, nucleotides, and vitamins. A growth medium that contains these materials, along with the items contained in minimal medium, is referred to as a **rich medium** or a **broth.** Broths, also known as **undefined media** ("media" is the plural form of "medium"), cannot be prepared by mixing together measured amounts of chemicals. Instead, they are prepared by **hydrolyzing,** or pre-digesting, organic materials such as meat or yeast. *E. coli* cells grow more rapidly in a broth, doubling in number approximately every 20 minutes.

E. Coli Cultures Grow in Two Forms of Medium

E. coli cells can be grown either **in suspension** or by **plating.** Cells grown in suspension grow in a liquid form of the growth medium, usually in a wide flask with a narrow neck and mouth. Suspension cultures generally appear as a turbid, brownish yellowish liquid (Figure 7.4a) and must be constantly shaken or swirled, both to prevent the cells from settling to the bottom of the flask and to increase the oxygen available to the cells in suspension. In contrast, cells grown by plating are placed on the surface of a gel form of the growth medium, prepared by heating the liquid medium with agar. The agar and growth medium mixture, spread across the bottom of a shallow, flat, circular **petri dish,** cools and hardens to a gelatin-like consistency. Plated cells grow as a **lawn**—a densely packed solid layer of rapidly dividing bacterial cells—on top of the agar growth medium. When phage-transfected *E. coli* cells are plated to form a lawn, the lawn develops plaques, the clear or semi-clear round holes caused by phage-induced lysis of some of the bacterial cells which make up the lawn. Plated cells can also be grown to form individual round groups of cells, called **colonies,** by sharply decreasing the number of cells placed in any one area of the plate (Figure 7.4b). Under these conditions, each colony represents all of the offspring of a single cell and is therefore composed of a number of genetically identical cells. In other words, each colony of plated cells represents a clonal population.

E. Coli Cultures Grow in Four Distinct Phases

The growth of bacterial cell cultures can be divided into four distinct phases, based on the rate of cell reproduction at various times after being placed in fresh growth medium (Figure 7.5). During phase one, called the **lag phase,** the cells divide slowly if at all. This period of slow or non-growth allows the cells to adapt

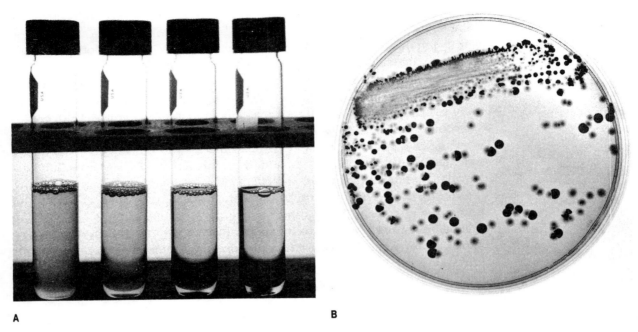

Figure 7.4
E. coli cells can be grown (a) in a suspension in which the presence of increasing numbers of bacterial cells causes turbidity in the medium growth, or (b) as individual colonies on a petri dish.

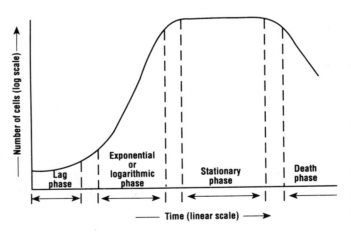

Figure 7.5
The growth curve of bacterial cell cultures can be divided into four stages, each of which is characterized by the rate of cell division.

to their new medium and to begin synthesizing the molecules and compounds they will require during the following period of active growth. As cells begin to make efficient use of the nutrients in the medium, they enter a phase of rapid cell division referred to as the **exponential** or **logarithmic (log) phase.** During this phase the number of cells doubles at its maximum rate. Eventually the cell culture reaches a point where the nutrients and/or oxygen in the growth medium become insufficient to support the rapid rate of exponential growth. At this point the number of cell deaths approximately equals the number of new cells being produced—this is called the **stationary phase.** The stationary phase comes to an end as the nutrients contained in the medium are depleted. As more cells die than are produced, the population enters the **death phase.** Cells can be maintained in the exponential phase almost indefinitely by providing the culture with a fresh infusion of medium before the culture ever reaches the stationary phase.

Making Competent Cells

Conversion of naturally incompetent *E. coli* cells to a competent state can be accomplished by treating the cells with a solution of the chemical compound **calcium chloride** in a **hypotonic buffer**—a liquid which contains fewer dissolved solids than does the cell which it surrounds. Since water always tends to move away from a hypotonic area, the cells swell in response to the calcium chloride solution as water leaves the hypotonic buffer and enters the cell. This treatment combined with **heat shock**—a brief period during which the temperature of the cells and their surrounding fluid is raised from 37°C (98°F) to 42°C (107°F)—causes the cells to become competent. The newly competent cells can then take up added **exogenous** or foreign DNA. This treatment of calcium chloride and heat shock, while easy, is an inefficient procedure and generally results in only 0.1% or less of the treated cells actually taking up and retaining exogenous donor DNA. Those cells that do take up and retain exogenous DNA are said to undergo **transformation** or to have become **transformed.**

The use of phage particles as cloning vectors simplifies the matter of DNA uptake somewhat since these vectors can enter a host cell by the relatively simple process of infection (Chapter Six). Transformation of bacterial cells with DNA derived from phage particles, often called **transfection,** generally results in approximately a ten-fold higher transformation efficiency than the calcium chloride/heat shock technique. This means that as many as 1% of the potential host cells may actually contain the exogenous target DNA after transfection. Transfection has the added feature of potentially resulting in **transduction:** the transfer of DNA from one bacterium to another via the release of infective phage particles. Although transduction, transfection and transformation have different technical meanings, the terms are often used interchangeably.

DETECTING TRANSFORMED CELLS

Transformed Cells Can Be Identified in Selective Media

Once an *E. coli* host cell population has been treated with both calcium chloride and a brief heat shock, it becomes competent and may take up and retain an exogenous DNA molecule or molecules under the right conditions. Cellular competency does not, however, guarantee that any particular cell will actually become transformed. As we have seen, only 0.1% to 1.0% of the host cells actually go on to become transformed cells. Thus, the next step involved in library production involves the task of identifying and singling out those relatively few transformed cells from the 99% of cells that did not take up and retain any exogenous DNA. The cloning of any donor DNA fragment using any vector molecule ultimately relies on the ability of the scientist to detect the presence of the vector once it has entered the host cell. If we are unable to identify the transformed cells, we certainly can't work with them!

One could complete this task of identification by examining individual cells one at a time. When we consider the billions of cells involved in even the simplest of laboratory procedures, we realize the impracticality of such a method. Instead, scientists have developed identification techniques that make use of **selective media**—growth media formulated to allow the growth of some cells while causing the deaths of other cells. In other words, selective media select the growth of some cell types at the expense of other cell types.

Scientists choose a particular selective medium for an experiment because it interacts in some way with a selectable marker located on the cloning vector (Figure 7.6). Suppose that we choose an *E. coli* host cell strain that is sensitive to the effects of the antibiotic tetracycline. The shorthand way of describing the phenotype of this host cell is **tet^S**. Because of their tet^S phenotype, untransformed host cells cannot survive in a selective medium containing tetracycline. In addition, suppose that we choose a **tet^R** cloning vector—one that carries the gene for tetracycline resistance as a selectable marker. Host cells, transformed with the tet^R vector, have incorporated the cloning vector molecule along with its gene for resistance to tetracycline. Therefore, these transformed cells have acquired the ability to survive in a selective medium containing this antibiotic. In other words, the phenotype of the host cells converts from tet^S to tet^R upon transformation. Transformed cells can be distinguished from untransformed cells by placing all of the cells in the selective medium. Only the transformed cells will survive and multiply. The medium will have **selected** only those cells which, through the acquisition of a cloning vector, are tet^R and thus able to grow in the presence of tetracycline.

The First Use of Selective Media as a Screen for Transformation

Production of a recombinant DNA molecule that could replicate in a host cell was first successfully carried out by Herbert Boyer and Stanley Cohen and their collaborators at Stanford and the University of California, San Francisco, in early 1973. These researchers used a small plasmid, originally isolated from *E. coli* bacterial cells and called pSC101 (pSC = plasmid of Stanley Cohen), as their cloning vector. This plasmid contains a number of useful features including a gene encoding tetracycline resistance and a unique *Eco*RI recognition site so that treatment of the circular plasmid with this enzyme yields one linear molecule for each treated plasmid. After linearizing the cloning vector, Boyer and Cohen took a preparation of a second bacterial plasmid which contained both a gene for resistance to the antibiotic kanamycin and an *Eco*RI restriction site. They also treated this second plasmid with *Eco*R1 to cleave it into a linear mol-

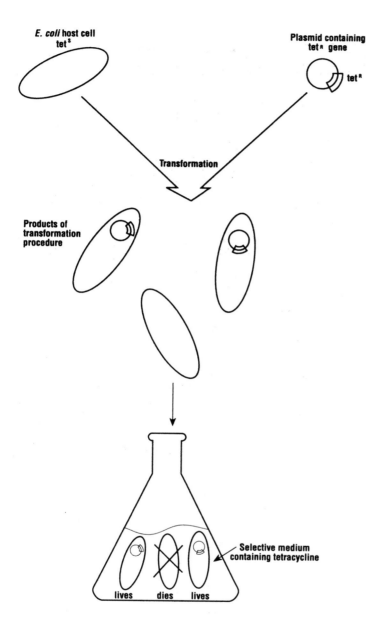

Figure 7.6
Selective medium is used to identify a subset of cells in a heterogeneous population. In this illustration tet^s host cells are transformed with a tet^r plasmid. The resulting population of treated host cells contains both successful transformants and cells that did not take up and retain the tet^r plasmid. All treated cells are grown in a growth medium containing tetracycline. This medium selects for successfully transformed host cells by preventing the growth of unsuccessful transformants.

ecule which would then be used as donor DNA. When they mixed the linear cloning vector plasmid DNA with the *Eco*RI-generated donor plasmid DNA, and DNA ligase was added, new recombinant circular plasmids were generated (Figure 7.7). Each of these new plasmids contained one or more *Eco*RI-generated linear molecules inserted into the *Eco*RI site of pSC101. These recombinant plasmids were then used to transform tet^{s}kan^s *E. coli* cells which were, in turn, plated on selective growth medium containing both tetracycline and kanamycin. Any bacterial cells able to grow in these conditions obviously had incorporated a recombinant plasmid, since untransformed host cells perished on this selective medium due to the lethal effects of both antibiotics. Thus, the selective medium promoted the growth of successfully transformed host cells, and discouraged the growth of untransformed cells.

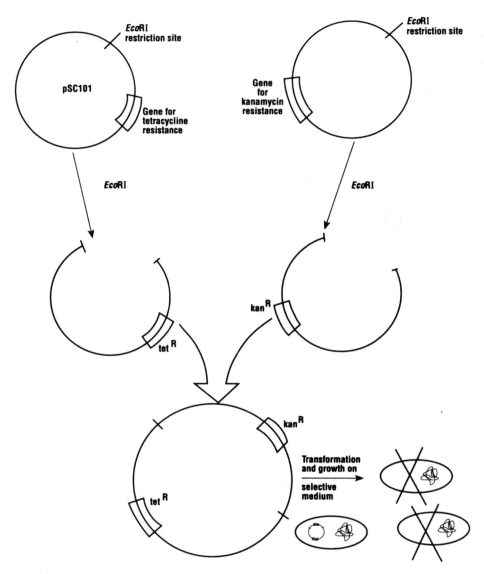

Figure 7.7
Schematic diagram of Boyer and Cohen's first cloning experiment.

IDENTIFYING RECOMBINANT VECTORS

Distinguishing Recombinant Vectors From Non-Recombinant Vectors

In many cloning procedures it is beneficial to distinguish transformed cells containing insert/vector recombinant molecules from those transformed cells with non-recombinant cloning vectors. This distinction can be made either during the construction of the recombinant vector or during the procedures which select for successful transformants.

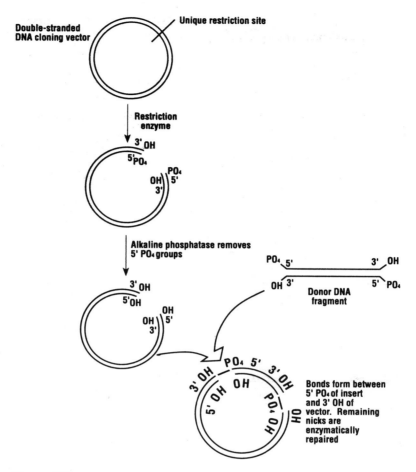

Figure 7.8
Ligation of DNA fragments requires the presence of phosphate groups at each of the 5' ends of the involved molecules. Removal of the 5' phosphate groups from any one fragment renders that fragment unable to circularize by self-ligation. Scientists exploit this fact by removing the 5' phosphate groups from plasmid molecules so that the cleaved plasmids are unable to recircularize without incorporating a DNA insert to provide the missing phosphate groups.

Alkaline Phosphatase Aids Construction of Recombinant Vectors

Construction of a recombinant molecule, consisting of a donor DNA fragment inserted into a circular cloning vector, requires the action of restriction endonucleases. These enzymes can create single-stranded sticky ends on DNA fragments which often bond to one another due to the complementary nature of the single-stranded bases. Successful repair or ligation of the broken sticky ends requires the presence of a chemical group called a **phosphate group (-PO$_4$)** on the 5' end of each of the two strands which make up the DNA fragment (Figure 7.8). The 3' ends terminate in a **hydroxyl (-OH)** chemical group. In order for the sticky ends to fully match up and ligate, each 3' hydroxyl group must bond to one 5' phosphate group. If the phosphate groups are not in place, ligation cannot occur. Scientists exploit this fact by using an enzyme called **alkaline phosphatase** to remove the phosphate groups from both of the 5' ends of the cleaved vector molecule. Once treated with this enzyme, the sticky ends of the vector can no longer ligate to one another and recircularize the vector since there are no 5' phosphate groups to link up with the 3' hydroxyl groups on the ends of the linear vector molecule.

Donor DNA fragments which are to be inserted into the cloning vectors are not treated with alkaline phosphatase. As a result, the 5' terminal phosphate groups remain in place on the insert molecules. When the untreated inserts are mixed with a preparation of the treated cloning vector molecules, the only 5' phosphate groups available to ligate to the vector-derived 3' hydroxyl groups come from the insert. Therefore, the only successful ligation events which can take place must involve both a vector and an insert.

Insertional Inactivation Detects Recombinant Vectors

An alternative to the method described in the preceding paragraphs involves a selection technique which simultaneously detects both the presence of a transformed cell and the presence of a DNA insert in the cloning vector of the transformed cell. This method, known as **insertional inactivation,** refers to the loss of a selectable marker from the vector due to the insertion of an exogenous DNA fragment into the gene which encodes that marker (See Chapter Six).

Insertional Inactivation Can Involve a Color Change

One application of insertional inactivation involves a visually detected color change. This system is based on the ability of the enzyme **beta-galactosidase** to cleave a colorless compound whose chemical name is **5-chloro-4-bromo-3-indolyl-beta-d-galactoside,** commonly known as **X-gal.** When X-gal is cleaved by beta-galactosidase it produces a sugar molecule called **galactose** and a bright blue, non-diffusible, indigo dye, easily detected on an agar plate as a round blue spot.

The gene which encodes beta-galactosidase can be included in a cloning vector which can, in turn, be used to transform a bacterial host cell. Any transformed bacterial colonies which grow on an agar plate in the presence of X-gal will be blue due to the action of the beta-galactosidase enzyme. If, however, a piece of exogenous DNA has been inserted into a restriction site within the gene which encodes beta-galactosidase, the gene will become non-functional. As a result, no enzyme will be produced to cleave X-gal and yield the blue indigo dye. Any bacterial colonies resulting from a transformation event involving this recombinant plasmid will be white. Simple visual examination can detect such colonies in the midst of a series of blue colonies. This technique differs from those involving selective media in that it allows the growth of both transformed and untransformed cells, while properly chosen selective media allow the growth of transformed cells alone. Neither of these techniques, however, offers aid in the identification of a particular target gene insert.

ANALYZING GENE LIBRARIES

Gene Libraries Must Be Screened

All types of libraries—cDNA and genomic, expressed and non-expressed—similarly contain a wide variety of donor DNA inserts which must be screened in order to locate any one specific target gene. Screening a population differs from the selection processes described earlier in this chapter in that selection involves identifying the entire subset of cells which have become transformed, while screening refers to the identification of particular sequences of donor genetic material within transformed cells.

Gene Libraries Don't Have Proper Indexes

Like the public library in your town, a gene library is a wondrous thing—with a very major difference. Your gene library, unlike your public library, has neither a friendly librarian nor a card catalog to assist you. To find a particular gene in your gene library—call it Gene A—you can't just go to the card catalog and look under the "A" listings. Things are much more confusing in the gene library: everything contained in the library is mixed together in an arbitrary fashion, so each thing must be individually examined to see if it is what you are looking for. It compares to an earthquake knocking over all of the shelves in your library just before you had to do research for your history paper—you would be faced with an enormous stack of books all jumbled this way and that. You would have to take each book from the pile, one at a time, and see if it contained the information you needed. Because all of the books would have been knocked out of place, the index of the card catalog would no longer be useful.

A gene library never has an index. Each entry in the library must be examined individually to determine its usefulness. The particular way that you choose to look at each entrant—to **screen** the library—depends on precisely what you are seeking.

Let's say that you have to find a particular stranger in a crowded train station. To succeed in your search at all, you must have some information about the stranger. Perhaps you know that he wears a red carnation in his lapel or that he speaks in a high falsetto voice. Maybe he carries a large green parrot on his shoulder at all times. You will proceed according to which piece of information you possess. If he wears a carnation, you will visually scan the crowd for the presence of that particular flower. If, on the other hand, you know about the falsetto, you will perk up your ears until you recognize such a voice.

Similarly, when you screen a gene library you must have some information. You may know the mRNA sequence which is complementary to your target gene, or perhaps you are aware of the identity of the polypeptide product encoded by the target gene. Maybe you have a piece of cDNA which you know represents the coding exons of the target gene. You will choose one of two basic screening possibilities depending on which information you have: you can screen for the target DNA sequence itself or, in the case of expression libraries, for the polypeptide product of the target DNA.

In any case you will use a **probe** to screen the library for your target. Simply put, a probe is a molecule that is able to recognize a particular target sequence. Probes can consist of cDNA, genomic DNA, RNA sequences, or even special protein molecules called **antibodies** (see below). The specific type of probe you use depends on the nature of your information about the target gene. If you know the mRNA, cDNA, or genomic DNA sequence, you will use a nucleic acid probe. Similarly, if you know the identity of the polypeptide product you will use an antibody probe which binds specifically to that molecule. Before we use our chosen probe to screen our growing library we will make a back-up copy of the library.

Replica Plating "Photocopies" Bacterial Colonies

The *E. coli* cells which contain the library, whether cDNA or genomic in origin, are often grown as colonies on petri dishes containing gelled growth medium. Ideally, each colony in the library represents a clonal population of cells and thus contains a single type of DNA insert. A library is useful only insofar as we can identify and manipulate the specific desired insert. Therefore we must examine, or screen, each colony in order to determine which DNA insert contains our target sequence. In order not to destroy the original library plate with this examination we use a technique called **replica plating** in which a piece of sterile velvet is carefully placed on the surface of the original library plate. Cells from each colony transfer and stick to the velvet which is then removed and carefully placed onto a new plate containing growth medium. Once again cells transfer, this time from the velvet to the surface of the medium on the new plate. As the cells grow, they form a new set of colonies which replicate the original plate. In this way, the cells of the replica plates can be examined while the original plate holds a back-up population of cells undamaged in any way by the experimental examinations.

All Screening Procedures Are Basically Similar

A number of protocols are used to screen genetic libraries. Although each protocol is unique, all protocols follow similar basic procedures (Figure 7.9):

1. Some material from the colonies or plaques—phage particles, whole cells, nucleic acids, or proteins—is transferred from the colonies to a piece of **nitrocellulose** filter paper. Nitrocellulose is a type of cellulose which has been treated with nitrates prior to being pressed into sheets of filter paper. The materials are carefully transferred so that they retain the same spatial relationship to one another as do the materials on the original plate.

2. The transferred material is treated to make it available for analysis. For example, if whole cells are transferred they are treated with a dilute basic solution, often sodium hydroxide, which both lyses the cells and causes the double-stranded DNA molecules to become single stranded.

3. A radioactive probe, complementary to the target molecule, is prepared. For example, radioactive cDNA may be prepared from a template of mRNA transcribed from a target gene.

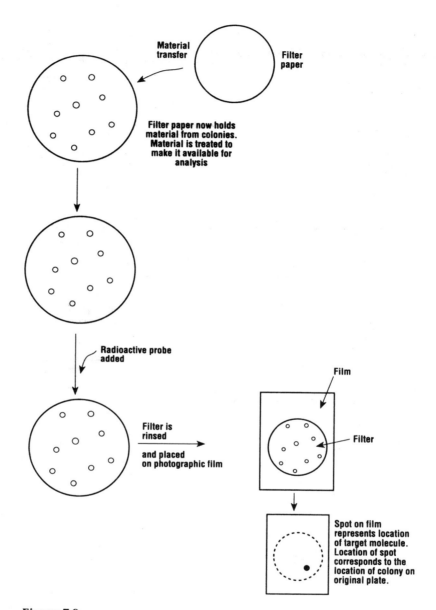

Figure 7.9
As described in the text, this figure shows an overview of a generalized library screening procedure.

4. The filter is then treated under conditions which cause the probe to bind specifically—or **hybridize**—to the target molecule. Under extremely restrictive conditions—known as conditions of **high stringency**—hybridization occurs only between two perfectly complementary sequences. As conditions of stringency are lowered, partially complementary sequences also hybridize with one another.

5. Excess probe material is washed away from the filter, which is then dried and covered with a sheet of unexposed photographic film for analysis by autoradiography. Any radioactive areas on the filter cause the film to become exposed in the immediate area surrounding the labelled probe. The location of the labelled probe on the filter corresponds to the location of the target molecule, and therefore the target colony, on the original petri dish.

PROBES FOR SCREENING

Let's talk about the different types of probes which can be used to screen a library and the type of target information they require.

cDNA as a Probe

Even if 97% of the protein produced by a given cell is the desired protein, that cell also produces small amounts of housekeeping proteins, if only to maintain the processes necessary to the life of the cell. Therefore, if we isolate the total mRNA from that cell, at least 3% of the mRNA will not be complementary to the gene which is of interest to us. In most cases the level of contamination is much higher than that. In fact, the average eukaryotic cell contains as many as 10,000 different types of mRNA molecules. Since we use an mRNA preparation as a template for the synthesis of a cDNA library, the library also contains some detectable level of DNA sequences which do not represent the target gene.

One type of probe used to screen such a library, a cDNA probe, can be made as follows: plasmids are isolated, along with their cDNA inserts, from a clonal population of transformed cells. The isolated plasmid DNA is then denatured so that the double-stranded molecules separate into single strands which are then attached to sheets of nitrocellulose paper. A mixture of mRNA molecules, complementary to many different gene sequences, can then be passed through the filter paper containing the population of homogeneous cDNA molecules isolated from the clonal population. Any mRNA molecule coming in contact with a complementary cDNA sequence sticks to the cDNA by complementary binding. All other mRNA molecules are washed away.

The mRNA molecules which remain bound to the cDNA on any one filter represent a homogeneous population because they are selected by their complementarity to the cDNA molecules. These cDNA molecules are also homogeneous because they are derived from a clonal population, all of whose members are genetically identical.

The bound mRNA can then be removed from the filters and added to a cell-free system that translates mRNA into protein molecules which can then be identified. This, in turn, identifies the cDNA insert in the original clonal population. In other words, if protein A is produced from the mRNA which is isolated from the filter, then the cDNA on the filter must represent gene A. Since the cDNA came from a clonal population, and since all of the cDNA in that population must be identical, the remaining cells of the original population must also carry plasmids which carry the cDNA gene for protein A.

Once the cDNA has been identified in this way it is called a **cDNA probe** and can be used to probe a population for a particular DNA or RNA sequence. This cDNA probe can also be used to rapidly screen nitrocellulose filters for the presence of its target gene sequence. For example, once we have a probe for gene A, that probe can be used to screen nitrocellulose filters produced from the clonal populations of any library to identify either gene A or sequences similar to gene A.

In some instances, a cDNA probe can be prepared directly from an mRNA whose identity is known. This happens only when a cell type exists whose natural mRNA population is predominantly of a single known type. An example of such an instance would be the use of **reticulocyte** mRNA as a template for a globin cDNA probe: a reticulocyte is an immature red blood cell whose mRNA consists predominantly of sequences complementary to the globin gene.

Production of Synthetic Oligonucleotide Probes

Sometimes the amino acid sequence of all or, more often, part of the target protein is known. Using this type of information and knowing the genetic code, the scientist can then predict a sequence of nucleotides which is complementary to that part of the target gene responsible for the synthesis of the target protein. This sequence of nucleotides, called an **oligonucleotide,** can then be built, radioactively labelled in the laboratory, and used to probe either genomic or cDNA libraries. Any cDNA sequence that binds to the oligonucleotide probe either represents the gene responsible for encoding the original target protein, or it represents a gene with a very similar structure.

Probes Can Be Used to Examine Uncloned Genomic DNA

Once a scientist isolates an mRNA or a cDNA probe whose identity is known, the probe can also be used to analyze genomic DNA sequences without first cloning them. This technique, developed in the mid 1970s by Dr. E. M. Southern, has come to be known as **Southern blotting** (Figure 7.10). In the Southern blotting technique, genomic DNA is cleaved with one or more restriction endonucleases. The resulting DNA fragments are then separated by agarose gel electrophoresis (Chapter Four). The double-stranded DNA fragments, now distributed throughout the electrophoresis gel, are denatured by soaking the gel in an alkaline solution of sodium hydroxide. As a result, the paired strands separate from one another to form single-stranded DNA fragments. After being brought back to a neutral pH, the entire gel, containing the bands of size-sorted denatured DNA molecules, is placed directly against a solid support. This support is usually a piece of filter paper which acts as a wick to draw a salt solution into the gel. Stacked on top of the gel is a sheet of nitrocellulose paper which is topped by a stack of absorbent material, usually in the form of paper towels, which serve to draw the salt solution up through the stack. As the salt solution passes through the gel matrix and into the neighboring nitrocellulose filter paper, it brings the single-stranded DNA molecules along with it. This process results in the transfer of a replica of the agarose gel to the nitrocellulose. Meanwhile an identified nucleic acid probe—a probe for the human hemoglobin gene, for example—is prepared so that it contains radioactive molecules. This labelled or **hot probe,** which can be RNA, cDNA, or even cloned genomic DNA, is placed in the solution which surrounds the nitrocellulose filter containing the replica

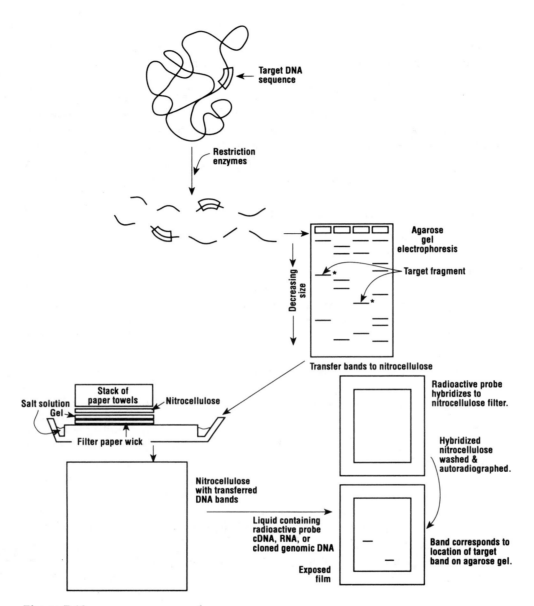

Figure 7.10
Southern blot analysis involves cleavage of DNA molecules with one or more restriction enzymes and separation of the resultant fragments by agarose gel electrophoresis. The DNA in the bands is transferred from the gel to a piece of nitrocellulose filter paper by allowing a salt buffer to flow through the gel and into the nitrocellulose. The nitrocellulose filter, which now holds a replica of the gel banding pattern, is treated with a solution containing radioactive probe molecules complementary to the target DNA fragment. After hybridization occurs between the probe and the target, excess radioactive material is rinsed from the filter, and the filter is subjected to autoradiography. The location of the band or bands which develop on the autoradiogram corresponds to the location of the target DNA fragment on the original agarose gel.

of the agarose gel. Under appropriately stringent conditions the probe then hybridizes to any DNA band containing a sequence which is complementary to the probe itself. Any radioactive probe/band complexes can be detected by the technique of autoradiography. As described earlier, this technique involves placing the treated nitrocellulose filter on a piece of undeveloped photographic film. Any place that the film and a radioactive label come into close proximity, the film develops. By looking at the location of developed spots on the film, we can determine the location of hybridized bands on the nitrocellulose. Since the nitrocellulose filter contains a direct replica of the original gel, we can determine the precise bands which contain a sequence of bases complementary to our radioactively labelled probe and thus identify a DNA region which contains a gene of interest.

Probes Can also Be Used to Examine RNA Molecules

A technique that parallels the Southern blotting technique described above has also been developed to probe RNA molecules. In this technique, known as **Northern blotting,** single-stranded RNA—either total RNA or purified mRNA—is size-separated by agarose gel electrophoresis on a gel containing a denaturing agent. The denaturant functions to prevent hybridization of the mRNA molecules amongst themselves during the electrophoresis procedure. The electrophoretic bands are then transferred to nitrocellulose filter paper which is then treated with a radioactively labelled probe of known identity, washed, and subjected to autoradiography in much the same manner as occurs with Southern blotting. Any bands on the newly exposed film represent bands on the gel which contain RNA molecules complementary to the known probe.

Expression Vectors Allow Direct Detection of Gene Products

Sometimes one can identify the presence of a eukaryotic gene in a bacterial host simply by looking for a eukaryotic gene product among the prokaryotic gene products normally produced by that host. This technique requires both an expression library and that the cloning vector which was originally used was an expression vector—a vector containing all of the regulatory elements necessary for its inserted gene to be expressed (Chapter Eight). When the inserted gene has been transcribed and translated by the host cell, it is often possible to detect the foreign polypeptide using immunological responses.

Cell Products of the Immune System Can Aid in Direct Detection

Our world is a dangerous place. Not only do obvious dangers—earthquakes and car accidents—wait to claim our lives, but other, more subtle dangers lurk. A myriad of microorganisms, including bacteria, viruses, and parasites, surround us in countless numbers, some of which are capable of instigating life-threatening illnesses. Why, then, do the vast majority of us live out normal life spans of seventy or eighty or more years with nothing more serious in the way of illness than a few colds or bouts of the flu? The answer to this question lies in the body's

immune system. The immune system is composed of a series of different types of cells which interact with one another to patrol the tissues of the body and protect it from the multiple invaders that try to interrupt its security.

A primary player in the role of the immune surveillance is a type of white blood cell which has the ability to produce **antibodies.** Antibodies, protein molecules able to bind to **antigens**—invading foreign cells or molecules—mark the antigens for destruction by other immune system cells. The system works something like that of loggers in a forest. Some loggers have the job of surveying the trees and marking, usually with a spray of paint, those which can safely be removed. Later, other workers come along and actually remove those marked trees, leaving unmarked trees in place. Antibodies play the role of the spray paint in the immune system: first an antibody binds to an invading antigenic entity. Then, other cells come along and destroy those entities which are marked by the antibody, ignoring unmarked cells and molecules.

The immune system functions smoothly and efficiently because of the **specificity** of binding between the antibody and the antigen: each particular type of antibody can only bind to a single type of antigen. In other words, if your body contains an antibody that binds to a particular antigen—call it antigen A—then that antibody will only bind to antigen A. Your body may also contain an antibody that only binds to antigen B. If that is the case, the anti-B antibody will never bind to antigen A, and the anti-A antibody will never bind to antigen B.

The specificity of antigen/antibody binding plays an important role in the direct detection of the product of a gene which has been inserted, via a cloning vector, into a host cell. The technique involved, often called **Western blotting,** bears some similarities to Southern and Northern blotting. (Note that although Dr. Southern invented the original Southern blot technique, there is no Dr. Northern or Dr. Western! These names are merely variations of the original developer's name, just as the techniques are simply variations of the original Southern blot technique.)

Western blotting differs from Southern blotting in that it uses a labelled antibody to probe proteins on a filter rather than using radioactively labelled nucleic acids to probe DNA or RNA sequences on a filter.

The basic technique for Western blotting involves a host cell, which is actively synthesizing a product from the inserted exogenous gene, and an antibody which binds specifically to that product. The host cells are grown by being plated on the appropriate growth medium. Exact replicas of the colonies on the agar plates are then made by replica plating.

In Western blotting, one of the replica plates is treated to release the product of the inserted gene. This treatment may involve lysis of the host cells if the product is not secreted by the cells, cleavage from a precursor form of the final product, or another similar procedure. In any case, a filter treated with the appropriate antibody—one that binds specifically to the target gene product—is placed over the treated replica plate. Any target molecules bind specifically to the antibodies on the filter paper. The resulting antigen/antibody complex on the filter paper then reflects the position of the original colony which produced the target protein. The filter paper can then be incubated with a radioactively labelled second antibody, specific for a portion of the first antibody. Once excess unbound labelled antibody is washed away, the filter can be subjected to autoradiography. The position of the exposed spots on the photographic film directly corresponds to the position of the target colonies on the master plate.

GENE LIBRARIES ARE REUSABLE

A gene library, either genomic or cDNA, can be maintained easily by providing sufficient fresh growth medium to the growing host cells. In addition, transformed host cells can be kept for long periods of time by freeze drying in the process of **lyophilization**. Using this technique, some strains of *E. coli* host cells have been kept alive for more than a quarter of a century!

Host cells can also be stored at temperatures of −70°C or in ultra-cold liquid nitrogen. Cells to be stored by freezing are placed in a growth medium, supplemented with either glycerol or dimethylsulfoxide (DMSO). This supplementation decreases the bursting of the cells due to ice crystal formation during the freezing process. Once frozen, the cells are quite stable and can be stored for several years. When thawed, previously frozen cells behave just like their never-frozen descendants. In addition, host cells can be stored in a liquid growth medium at −20°C if the medium is supplemented with as much as 50% glycerol or DMSO. Cells stored in this state also maintain their viability for several years.

The ability to store host cells, and therefore libraries, for long periods of time, coupled with the ease with which most host cells are grown, means that a library can be used over and over again. Each use differs only in the choice of target gene and probe!

REVIEW AND SUMMARY

1. Once a recombinant vector has been synthesized, it must be inserted into an appropriate host cell in order to allow replication of the exogenous genetic material.

2. Cells which are able to take up exogenous DNA, including recombinant plasmids, are called **competent** cells. Some cell types are naturally competent while other cell types are not. *E. coli* cells must be treated with calcium chloride and heat shock, or a similar procedure to induce a competent state. Cells which take up recombinant plasmid DNA are called **transformed** cells. The efficiency of transformation is generally extremely low and ranges from 0.1 to 1.0%

3. *E. coli* cells are often chosen as host cells for a number of reasons: the *E. coli* chromosome has been well characterized and many of its polypeptide products identified. *E. coli* cells, easy to grow in the laboratory, have a short generation time which typically ranges from 20 to 60 minutes, depending on the growth medium in use.

4. Transformed cells can often be identified with the use of **selective media** which interacts in some way with a selectable marker located on the cloning vector. Typically the cloning vector carries a gene which confers drug resistance on a transformed cell, thus allowing only transformed cells to grow in selective media containing that particular drug. Non-transformed cells cannot grow in the selective medium.

5. The presence of an inserted donor gene in the cloning vector can sometimes be determined by **insertional inactivation.** This term describes the loss of a gene product due to insertion of foreign DNA in a recognition sequence located within a particular gene. Examples of insertional inactivation involve both variations in drug resistance phenotype and in color phenotype.

6. Treatment of a linearized cloning vector with **alkaline phosphatase** removes the 5′ phosphate groups, necessary for the religation of the vector's cleaved ends. Thus, pre-treatment of the cleaved vector with this enzyme ensures that only recombinant vectors result when a preparation of cleaved vector molecules is incubated with a preparation of donor DNA fragments in the presence of DNA ligase, because the ends of the treated vector will be unable to ligate to one another.

7. The two primary types of gene libraries—genomic and cDNA—differ from one another as follows: genomic libraries represent the entire donor genome while cDNA libraries represent only those DNA sequences being transcribed at the time of library preparation. Neither type of library possesses an index and must therefore be analyzed or screened with specific probe molecules to find and identify target sequences.

8. The basic screening technique includes the transfer of some component of the host cells or the phage vector to a solid substrate, usually nitrocellulose. Subsequent treatment of the substrate with a labelled probe which is complementary to the target sequence follows. Probes can be cDNA, genomic DNA, RNA, or antibody molecules. Specific probes are chosen based on information known about the target sequence. Probes will hybridize to complementary molecules on the substrate. Hybridized probe/target complexes are visualized with **autoradiography.** Alternatively, mRNA probes can be removed from the target molecule and used to direct protein synthesis in order to identify the original gene sequence.

9. **Southern blotting** is the analysis of DNA sequences with either a DNA or an RNA probe. **Northern blotting** is the analysis of RNA sequences with either a DNA or an RNA probe. **Western blotting** is the analysis of proteins with an antibody probe.

10. Transformed host cells can be maintained for long periods of time by storage at very cold temperatures, making gene libraries reusable. Frozen cells, thawed many years after their original storage, behave as though they had never been frozen. Thus, gene libraries can be probed, stored, regenerated, and reprobed many times.

Factors that govern gene expression

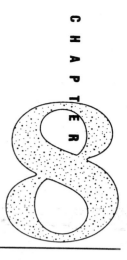

Jonah and Becky P. are two and four years old, respectively. This year on Becky's birthday their family acquired a baby kitten. The kitten, which the children imaginatively named "Kitty," had soft, pure white fur. Kitty also had icy blue eyes. The family enjoyed playing with their pet and watching her grow. As Kitty grew she developed black markings. Soon her legs were covered with soft black fur, as were her ears, tail, and face. The fur of her body, however, remained white (Figure 8.1).

Kitty, like all other Seal-point Siamese cats, is genotypically **albino.** This genotype causes them to be born with white fur and blue eyes. The gene responsible for the albino coloration, found in all of the cells of their bodies, has an unusual trait: its expression is **temperature dependent.** That means that the expression of the gene varies as the mean temperature of its immediate environment varies. In the case of Siamese cats, the gene that causes the white albino phenotype also has the ability to produce the dark pigment known as **melanin.** However, melanin is produced only in cells which exist at a relatively cool temperature; the extremities of cats—and humans, as well as of many other animals—are slightly cooler than the main body of the animal. This slightly cooler temperature allows the eventual production of melanin in these areas, resulting in a darker coloration of the extremities while the warmer areas of the body remain the initial unpigmented color.

OVERVIEW In matters concerning gene expression, as in matters concerning life in general, most issues cannot be resolved in black and white. Consequently, there is more to gene expression than saying a gene is either "**on**" or "**off**," "**expressed**" or "**unexpressed.**" The shades of gray come into play when we look at a gene whose expression is temperature sensitive as described above, or whose expression varies in a manner which is dependent on the expression of other genes, or is dependent on the function of other sequences of deoxyribonucleotide bases along the DNA molecule.

Figure 8.1
Siamese cats develop their characteristic markings with age.

This chapter concerns the control of gene expression. Many gene products are common to many different types of cells. These so-called **housekeeping gene** products, important to the basic existence of the cell, include such things as DNA repair enzymes and the enzymes necessary for DNA replication. Other gene products, however, are restricted to certain cell types. For example, only islet cells of the pancreas produce insulin, digestive enzymes form only in cells of the gastrointestinal tract, and growth hormone comes only from cells of the pituitary gland. Yet, despite this restriction in the gene products of a single cell type, all nucleated cells in a single organism have the same basic genetic makeup. This fact presents a dilemma: if the islet cell has genes to encode both insulin and growth hormone, why is only the former hormone produced by these cells?

Prokaryotic organisms, despite the absence of the cellular differentiation characteristic of multicellular eukaryotic organisms, also exhibit variable patterns of gene expression. As a prokaryotic cell goes through its life cycle, genes turn on and off in a regular pattern. In addition, the genes of prokaryotic cells are often expressed at varying levels in response to variations in the environment of the cell. How are these patterns established and controlled?

Discussions of genetic engineering and recombinant DNA techniques pose these important questions because these sciences often work toward the successful expression of a foreign donor gene by the enzymes of the host cell. What characteristics of the cloning vector or of the donor gene itself control the expression of this gene? In order to more fully understand this aspect of genetic engineering, let's consider the various control mechanisms responsible for turning some genes on and some genes off within a single cell—whether prokaryotic or eukaryotic—and answer the following questions:

1. What is a repressor and why should a particular gene be repressed?
2. How does the regulation of genes affect the phenotype of a cell and therefore an organism?
3. How do puffs relate to the concept of gene regulation? What about cat boxes?
4. How do prokaryotic and eukaryotic genes differ from one another? What about chromosomal structures?
5. Are the mechanisms of genetic control similar or different for prokaryotic and eukaryotic genes?

THE CONTROL OF PROKARYOTIC GENE EXPRESSION

Bacteria Respond to Rapidly Changing Environments

Bacterial cells generally live in environments subject to rapid changes in character. Consider the *E. coli* cells which inhabit your gastrointestinal tract. When you wake up in the morning after having not eaten all night, those *E. coli* cells are living in an environment relatively sparse in nutrients. Shortly after arising, however, you eat a hearty breakfast and drink a big glass of milk. Suddenly, your intestines bathe those bacterial cells in an environment extremely rich in **lactose,** the primary sugar found in milk. Lactose is a **disaccharide**—a sugar molecule composed of two subunits—which the enzyme **beta-galactosidase** breaks down into its **monosaccharide** subunits: **galactose** and **glucose** (Figure 8.2). The monosaccharide subunits can, in turn, generate energy which the *E. coli* bacterial cell uses to fuel the ongoing processes of growth and reproduction.

An *E. coli* cell surrounded by a lactose-free environment contains few if any molecules of the beta-galactosidase enzyme. When, however, the cell's environment suddenly receives lactose, things change. Almost immediately, the bacterial cells begin to synthesize the enzymes necessary to digest lactose into its subunit components. In fact, within minutes the *E. coli* cell contains thousands of molecules of beta-galactosidase along with equal numbers of two other types of enzyme molecules which aid the bacterial cells in breaking down the lactose disaccharide molecule (Figure 8.3). As the supply of lactose diminishes, these enzymes are no longer needed. Rather than waste energy producing enzymes which serve no immediate purpose, the *E. coli* cell quickly halts their production.

How does the *E. coli* cell so precisely control the synthesis of certain enzymes? Why are some bacterial enzymes produced rapidly in response to need, and not produced at all when they are not needed?

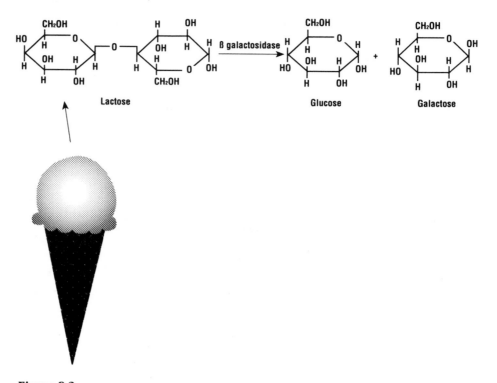

Figure 8.2
During digestion, the milk sugar lactose is broken down into glucose and galactose by the enzyme beta-galactosidase.

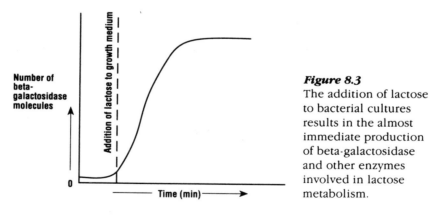

Figure 8.3
The addition of lactose to bacterial cultures results in the almost immediate production of beta-galactosidase and other enzymes involved in lactose metabolism.

Jacob and Monod Identify the *Lac* Operon

Although it seemed clear to scientists that the addition of lactose to the *E. coli* environment somehow induced the synthesis of the bacterial enzymes necessary to digest that disaccharide, it wasn't until 1961 that French geneticists Francois Jacob and Jacques Monod elucidated the mechanism of this induction process. Through a series of in-depth experiments, these researchers found five components, consisting of four distinct genetic regions and one regulatory molecule, which are involved in the complex mechanism governing the expression of the gene products required for the digestion of lactose. These components include:

1. The **regulator gene:** this gene encodes a protein which changes its molecular configuration in response to the presence of lactose.

2. The **repressor protein:** this molecule, produced by the regulator gene, changes its configuration by binding to lactose. When lactose is unavailable, the shape change is not induced, and the repressor molecule retains its initial configuration.

3. The **operator:** this base sequence is the region to which the repressor molecule binds when it is in its initial configuration. The operator sequence overlaps the **promoter sequence.**

4. The **promoter:** this DNA sequence provides the site to which RNA polymerase (Chapter Two) binds prior to the initiation of transcription of RNA molecules which are complementary to the **structural genes.** Neither the promoter nor the operator sequences are themselves transcribed.

5. The **structural genes:** these three genes encode the three enzymes involved in the digestion of lactose.

Taken together, these five components make up the ***lac* operon.** In this term "**lac**" refers to the lactose molecule, and "**operon**" refers to a group of functionally related prokaryotic genes which act in a coordinated fashion and encode the various enzymes of a single biochemical pathway along with their controlling sequences.

Figure 8.4 shows how the five components of the *lac* operon work together to provide the exquisitely sensitive control that *E. coli* cells have over the synthesis of the enzymes used to digest lactose. The regulator gene encodes the repressor molecule in a **constitutive** fashion: the regulatory gene is always "on" and repressor molecules are always present in the cell. When the environment of the bacterial cell contains little or no lactose, repressor molecules retain their initial configuration and bind to the operator sequence. As a result of the overlap between the promoter and the operator sequences, the RNA polymerase enzyme becomes physically unable to bind to the promoter sequence, so that the associated structural genes will be neither transcribed nor translated. Thus, the absence of lactose in the environment directly leads to a shut-down of the genes which encode the enzymes of the *lac* operon.

In contrast, lactose present in the environment of the bacterial cells binds to the repressor molecule, thus inducing a shape change which renders the repressor unable to bind to the operator. As a result, the RNA polymerase enzyme, no longer blocked from binding to the promoter, initiates transcription and translation of the associated structural genes to produce the enzymes needed to digest lactose. When all of the lactose has been digested the lactose-induced shape change no longer occurs, so that the repressor can once again bind to the operator. The newly lactose-free repressor molecule thus turns off expression of the three structural genes.

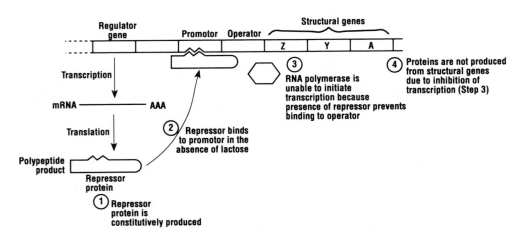

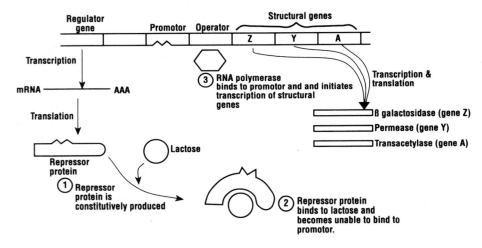

Figure 8.4
The *lac* operon regulates production of the three enzymes involved in the digestion of lactose. These enzymes are not synthesized unless the constitutively produced repressor protein is inactivated by the presence of lactose itself. *Lac* operon function is described in greater detail in the text.

In summary, the presence of lactose itself induces production of the enzymes that aid in the digestion of this sugar by blocking the action of the repressor molecule. When lactose is absent from the cell's environment, the repressor carries out its blocking function and blocks synthesis of the three involved enzymes by preventing RNA polymerase from binding to the promoter. For these reasons lactose is known as an **inducer** for the *lac* operon: the operon synthesizes its gene products only when its function is **induced** by the presence of lactose.

The *Lac* Operon Is a General Model

As research into the questions regarding prokaryotic gene control continued, scientists saw that the *lac* operon could serve as a model to describe the general mechanisms of bacterial gene regulation. In fact, research confirms that most bacterial enzyme-encoding genes are regulated in a fashion similar to that of the *lac* operon. Although most bacterial operons bear great similarity to one another, some significant differences exist. For example, the **trp** (pronounced "trip") **operon**, responsible for the synthesis of enzymes needed to convert certain raw materials into the amino acid **tryptophan**, is a **repressible operon**. Unlike the *lac* operon whose expression requires the presence of an inducer molecule, the *trp* operon functions continuously to synthesize a series of enzymes unless a **corepressor** molecule is present. The corepressor, in this case tryptophan itself, binds to the repressor molecule, which in turn to binds to the operator sequence. In this way, the presence of tryptophan represses the synthesis of the enzymes needed to produce that amino acid. Thus, while the structural genes of the *lac* operon are not expressed unless they are induced, the structural genes of the *trp* operon are always expressed unless they are repressed.

Some Promotors Are More Effective Than Other Promotors

As we have seen, the promoter of a prokaryotic operon is a site located close to the structural genes and to which the RNA polymerase enzyme binds. Once bound, the polymerase molecule proceeds along the length of the operon and initiates the transcription of the involved structural genes. The mRNA molecule produced by this process carries instructions for the translation of all of the genes found within a single operon. In other words, the mRNA molecule produced upon transcription of the *lac* operon contains information sufficient to encode the three separate enzymes whose genes are contained in this operon. Such a complex mRNA molecule is said to carry a **polycistronic** message.

Another interesting characteristic of prokaryotic cells involves the fact that their genetic material is not separated from the rest of the cell by a nuclear membrane. This lack of a nuclear membrane allows the enzymes of translation to have access to the mRNA molecules as it is being produced so that transcription and translation can proceed simultaneously. Such is not the case in a eukaryotic cell, where the mRNA is transcribed within the nucleus. The nuclear membrane separates newly synthesized mRNA molecule from the translation enzymes, so the mRNA must be transported through this membrane in order for translation to take place. Therefore, transcription and translation take place in two distinct stages in the eukaryotic cell.

The initiation of transcription requires the action of RNA polymerase. Therefore, the effectiveness with which an RNA polymerase molecule binds to the promoter of an operon affects the rate of transcription of the associated structural genes. In the mid-1970s researcher Dr. David Pribnow examined the base sequences of several bacterial and phage promoters to determine the identity of any sequences which appeared to affect the efficiency of RNA polymerase binding.

Dr. Pribnow found that, of the promoters he examined, all had two short nucleotide sequences which closely resembled one another. One sequence consisted of 6 base pairs and was centered approximately 10 base pairs **upstream**—away from the structural genes—of the **transcriptional start site:** the location at which transcription of the structural genes begins. The other sequence consisted of 10 base pairs which were centered about 35 base pairs upstream of the transcriptional start site. These sites have come to be called the **Pribnow box** and the **-35 sequence,** respectively.

Dr. Pribnow and his colleagues then determined the **consensus sequence** of these sites. A consensus sequence represents the specific base that is most likely found in each particular location within any specific sequence of bases. Examples of consensus sequences can be determined for many things including hard-to-spell words. For instance, if you ask ten people how to spell the ten-letter word "**antecedent,**" which means "going before or preceding," you will probably get several different responses. One person might say "**anticedent,**" while another person will guess "**antisedent,**" and yet another, "**antesedant.**" As you look at the various answers you see that nearly each one begins with "ant" for the letters in positions #1, #2 and #3. However, in position #4 six people said "**e**" and four chose "**i.**" The consensus for this position is therefore the letter "**e.**" In position #5 half of the people put a "**c**" while the other half chose "**s,**" the consensus therefore being "**c or s.**" Using this technique, a consensus sequence—here a consensus spelling—can be determined for all of the involved positions.

Although Dr. Pribnow's analyses of various prokaryotic organisms showed that the two stretches of nucleotides varied slightly from one organism to another, researchers could determine a consensus sequence representing the specific base most likely to appear in each particular location of the two sites. These consensus sequences are TTGACA (-35 sequence) and TATAAT (Pribnow Box). Furthermore, the promoters containing regions that more closely resembled the consensus sequences tended to be somewhat more efficient at binding RNA polymerase, while those promoters containing sequences that were less like the consensus sequences tended to be less efficient. Some promoters therefore apparently increase the efficiency of gene expression.

THE INTERACTION OF PROMOTERS AND CAP PROTEINS

CAP Proteins Are Involved in Positive Regulation

Positive regulation refers to situations in which the binding of an **activator** protein to the promoter increases the binding of RNA polymerase, and therefore also increases the rate of transcription of the involved structural genes. In other words, the activator protein enhances protein synthesis.

A category of bacterial proteins, called **catabolite activator proteins** or **CAP** proteins, can bind to a special DNA site—called the **CAP binding site**—in close proximity to an operon's promoter. This binding appears to enhance the binding of RNA polymerase to the promoter and thus enhances the transcription of the associated structural genes. In order for the CAP protein to bind most efficiently to the CAP binding site, the protein must first bind to a molecule called **cyclic adenosine monophosphate** or **cAMP.** The concentration of this intracellular chemical messenger is regulated by the various metabolic activities of the cell itself. Once this binding takes place, the CAP/cAMP complex molecule binds to the CAP binding site with relatively high efficiency. The presence of the CAP/cAMP complex then apparently stimulates the binding of RNA polymerase to the promoter site and thus enhances expression of the structural genes. The structural genes can therefore be said to be under positive control by the CAP/cAMP complex.

An example of the role of the CAP protein in positive regulation can be seen when an *E. coli* cell lacks its optimum food source, glucose. Without glucose, large amounts of the intracellular chemical messenger cAMP accumulate in the cell. A cAMP molecule then binds to a CAP protein molecule, and the resulting CAP/cAMP complex binds to the CAP binding site associated with various operons, all of which produce enzymes that break down sub-optimum food sources. These operons, including the *lac* operon, are therefore under positive control by the CAP protein which causes an increased level of gene expression as a result of a lack of available glucose.

In summary, the CAP protein functions in positive control situations by first binding to cAMP, after which the CAP/cAMP complex molecule binds to the CAP binding site. The binding of the CAP/cAMP complex to the CAP binding site increases the efficiency with which the RNA polymerase binds to the promoter and therefore increases the production of mRNA molecules transcribed from the associated structural genes.

CAP Proteins Are Also Involved in Negative Regulation

CAP proteins are also involved in some types of negative regulation. This type of negative regulation, which is often called **catabolite repression,** can be seen in the case of an *E. coli* cell which is provided with both the lactose disaccharide molecule and the glucose monosaccharide molecule as food sources. Although both sugars provide excellent sources of energy for the bacterial cell, glucose monosaccharide, the more efficient source, requires no expenditure of energy to break it down into its sugar subunits. In contrast, energy is used when the cell breaks down the lactose disaccharide to extract energy. For this reason, *E. coli* cells here use the catabolite repression mechanism which enables them to use glucose preferentially for energy even in the presence of lactose or other complex sugars. Let's look more closely at catabolic repression.

The presence of the monosaccharide glucose in the environment surrounding an *E. coli* cell induces two changes within the cell. First, it indirectly causes an alteration in the structure of the CAP protein. Secondly, it inhibits the accumulation of cAMP. As a result of these two events, the CAP protein binds to the CAP binding site with far lower efficiency and therefore does not greatly enhance the production of mRNA molecules from the structural genes of the involved operons. In practice, this means that the CAP protein does not efficiently bind to the CAP binding site of the *lac* operon and therefore that the bacterial cell does not waste energy transcribing the structural genes of the *lac* operon to produce the enzymes necessary to digest lactose when the more easily utilized glucose is available.

Prokaryotic Transcriptional Gene Control Is Complex

As described above, the *lac* operon produces a repressor protein and three different structural protein products in quantities which depend on the presence or absence of both lactose and glucose, as well as on the functioning of the *lac* repressor protein and the CAP protein. These factors interact with one another to fine-tune the responses of the *E. coli* cell to the specific nutrients present in its environment. As a result of these sensitive and complex interactions between the various controlling elements, the *E. coli* cell controls the processes of transcription to make the most efficient use of all available nutrients. Although most prokaryotic gene control takes place at the transcriptional level, other control mechanisms do exist. Let's take a look at a few of these alternative mechanisms.

PROKARYOTIC TRANSLATIONAL GENE CONTROL

So far in this chapter we have discussed the control of prokaryotic gene expression by controlling various aspects of the process of transcription. Alternative possibilities for gene control also exist at the level of translation: if mRNA is prevented from undergoing translation, the level of protein drops regardless of the level of mRNA. Similarly, speedy translation of a few mRNA molecules may yield more protein than very slow translation of many mRNA molecules.

The term **translational regulation** refers to the series of events that may limit the actual processes of initiation and elongation of the polypeptide on the ribosome/mRNA complex during translation. If either initiation or elongation is blocked, the quantity of the gene product will be reduced.

Ribosome Binding Site Is Discovered

In 1974 Australian researchers Dr. John Shine and Dr. Larry Dalgarno first described the existence of a specific area to which ribosomes bind on an mRNA molecule. This site, now known as the **Shine-Dalgarno sequence** or the **ribosome binding site,** appeared to act as an indicator of the correct position for the binding of an mRNA molecule to a ribosome. You recall from the discussions of translation in chapter two and of mutations in chapter three that correct position of the mRNA on the ribosome is imperative, as it establishes the reading frame of the entire genetic message which is to follow. The wrong reading frame can lead to a hopelessly garbled genetic message which, in turn, can lead to an incorrect polypeptide product.

The Shine-Dalgarno sequence, with a consensus sequence of AAGGAGGU, complements a short sequence of nucleotides located at the 3' end of one of the two **ribosomal RNA—rRNA—**subunits which make up a **ribosome—**the cellular organelle where protein synthesis occurs. The Shine-Dalgarno sequence consists of 3 to 9 bases and is located from 3 to 12 bases upstream of the **translational start site.** This site, marked by the AUG codon which specifies the amino acid methionine, marks the position at which the translation process begins. The complementarity which exists between the Shine-Dalgarno sequence and the ribosome probably accounts for the correct positioning of the mRNA molecule on the ribosome.

Researchers believe that interference with the Shine-Dalgarno sequence can decrease the efficiency of translation by inhibiting the correct placement of the mRNA molecule on the ribosome. We see an example of this in the synthesis of ribosomes themselves. Ribosomes are composed of two rRNA molecules and some associated ribosomal protein molecules, known as **r-proteins.** When the rate of synthesis of the protein molecules exceeds the rate of synthesis of the rRNA molecules, not all of the protein molecules will be incorporated into newly produced ribosomes. Instead, the excess r-protein molecules probably interact in some way with the Shine-Dalgarno sequence of the mRNA which encodes the ribosomal proteins themselves. Through this interaction, it is postulated, correct binding of the r-protein mRNA to an established ribosome is inhibited. In this way, synthesis of new r-protein molecules decreases until the excess ribosomal proteins are used up in the process of synthesis of new ribosomes. At this point, inhibitory binding of r-proteins to the Shine-Dalgarno sequence ends, and the level of r-protein synthesis increases once again.

The position of the Shine-Dalgarno sequence also affects the relative efficiency with which it directs the positioning of mRNA molecules on ribosomes. Although Shine-Dalgarno sequences apparently vary somewhat in their positioning, experiments show that this sequence functions optimally when it is centered 8 bases upstream of the translational start site. Moving the Shine-Dalgarno sequence either upstream or downstream of the optimum location decreases its efficiency, as does hiding it within the folds of an mRNA molecule.

THE CONTROL OF EUKARYOTIC GENE EXPRESSION

Eukaryotic Gene Control Is Also Complex

Although similarities, such as the existence of regulatory proteins, exist between the gene control mechanisms of eukaryotes and prokaryotes, they do exhibit many differences. In fact, much less is known about eukaryotic than about prokaryotic gene control, in part because of the greater complexity and size of eukaryotic genomes.

When we examine groups of prokaryotic cells and compare them with groups of eukaryotic cells in the form of an organism, a number of differences appear. In a group of prokaryotic cells, even in a multicellular colony, all of the cells are the same. This sharply contrasts with populations of eukaryotic cells as is evidenced by the existence of **differentiated cells** in the eukaryotic organism.

While we know that each eukaryotic organism, including each human being, develops from a single fertilized cell called a **zygote,** we also realize that each eukaryotic organism consists of many different types of cells which interact with one another to form a functioning being. For example, you began existence when a single ovum was fertilized by a single sperm. At that point, the resulting zygote contained the potential to divide and form all of the different types of cells which make up a human being. As that zygote divided to form an embryo that developed into a fetus and from there into a baby which then grew to be a full grown adult human being, trillions of cells were produced. All of those cells descended from that one fertilized egg cell. Although each of those cells—with the exception of red blood cells which contain neither nucleus nor genetic material—contains a closely similar genetic complement, many of those cells have specific functions which cannot be duplicated by other cells. For example, the cells of the retina of your eye allow you to process signals of light and dark, the cells of your pancreas produce insulin, and the cells of your heart continue to contract and cause the circulation of blood throughout your body during the entire course of your life.

Although all of the cells in this great variety of differentiated cell types contains identical genetic complements, not all of the genes function all of the time. Instead, each cell type expresses a particular subset of all of the genes which it contains. In some types of cells that subset remains constant for the life of that cell. In other types of cells that subset changes with the maturity of the cell. The important point: some genes in a single cell are expressed while other genes are not. In fact, estimates show that each type of differentiated cell uses less than ten percent of its total number of genes. That type of selective gene expression can only be the result of genetic control mechanisms.

Eukaryotic Regulation at the Level of Transcription

As with prokaryotic cells, the majority of genetic regulation in eukaryotic cells takes place at the level of transcription. Thus the control over the final polypeptide production is achieved through manipulating the mRNA molecules which are transcribed from the DNA templates. Both eukaryotic and prokaryotic transcriptional regulation mechanisms make use of regulatory proteins which are transcribed from genetic sequences known as regulator genes. Here the similarity ends, however.

We saw how most prokaryotic genes are organized into operons whose expression is governed by repressor proteins: the presence of the repressor protein halts transcription of the genes of the operon. Eukaryotic genes, on the other hand, are not organized into operons and most often appear to be regulated by **activator** proteins. This means that eukaryotic genes generally remain unexpressed unless the activator protein is present. The presence of activator proteins instead of repressor proteins may reflect the relative rates of transcription of eukaryotic and prokaryotic genes: most eukaryotic genes in a given cell are rarely, if ever, transcribed. Therefore, it is more efficient to turn genes on rather than to turn them off. Conversely, most prokaryotic genes are transcribed most of the time, so in this case it is more efficient to turn the genes off when necessary, rather than maintaining a supply of activator proteins to turn them on. It should be noted that a few prokaryotic operons do function in an inducible fashion. Such operons, which require an **activator** protein, are transcribed only in the presence of the activator molecule.

Eukaryotic DNA Contains Repetitive Sequences

Before we can compare the control of eukaryotic gene expression to the control of prokaryotic gene expression, we must discuss some of the many differences in eukaryotic and prokaryotic gene organization.

The eukaryotic and prokaryotic genomes obviously differ in that the former seems to have a great deal more DNA than the latter. For example, the prokaryotic *E. coli* cell has a mere 5.0×10^6 base pairs in its genome while a eukaryotic human cell contains approximately 3×10^9 base pairs—about six hundred times more than the *E. coli* genome. Although the human organism is clearly more complex than the *E. coli* organism, does this complexity account for the enormous difference in genome size? Many scientists think not. Perhaps an explanation for this discrepancy can be found in the relative rate of repetition in the DNA base sequences.

Much of the extra DNA of the eukaryotic cell can be found in the form of **repetitive DNA.** These DNA sequences, composed of 100 to 300 base pairs, may be found in many thousands of copies throughout an organism's genome. An example of a repetitive DNA sequence is the **Alu** sequence, so named because it carries within itself the recognition site for the restriction endonuclease *Alu*I. The human genome contains approximately three hundred thousand copies of the Alu sequence of repetitive DNA. In contrast, the 5 million base pairs that make up the *E. coli* genome contain no repetitive DNA sequences.

Eukaryotic *vs* Prokaryotic Gene Structure

A second feature distinguishes prokaryotic from eukaryotic genes: eukaryotic genes are composed of both introns and exons, while prokaryotic genes contain only exons. As discussed in chapters two and six, this feature makes eukaryotic genes comparatively larger than prokaryotic genes. The various processing steps necessary to form a functional polypeptide product from a gene containing both introns and exons provides many additional sites for possible mechanisms of genetic control. These sites, including the post-transcriptional modification of mRNA and the post-translational modification of the polypeptide product, may come into play in the mechanisms of eukaryotic gene control.

Eukaryotic Chromosomal Structure Is Complex

Eukaryotic chromosomes contain an enormous amount of DNA relative to their size and to the size of the nucleus which contains them. Estimates suggest that the DNA contained in the chromosomes of each human cell, stretched to its full length, would be more than a meter long! Yet the size of an average eukaryotic cell approximates only 25 microns in diameter, and the nucleus is even smaller than that. How is it that a meter length of DNA can be squeezed into a space that small?

The Higher Order Packing of Eukaryotic DNA

As discussed in chapter four, the strands of DNA which make up a chromosome are efficiently packed along with proteins known as **histones.** Specifically, each strand of eukaryotic DNA wraps around a series of histone proteins which act as spools. As a result of this wrapping process, each strand of eukaryotic DNA resembles beads on a necklace. Each bead of histone/DNA complex is called a **nucleosome.** Thus, a single strand of DNA is arranged into a series of nucleosomes. This nucleosome strand then supercoils on itself so that it forms a thick fiber—called a **chromatin fiber**—which is approximately three times thicker in diameter than the nucleosome strand from which it was formed. The next level of DNA packaging requires that the chromatin fiber be organized into **looped domains,** each of which contains a total of 20,000 to 100,000 base pairs. The looped domains, in turn, loop and fold onto themselves to form an extremely compact chromatin fiber, nearly 70 times thicker than the original beads-on-a-necklace nucleosome strand, and 350 times thicker than the original unpacked DNA strand itself (See Figure 4.5)! The chromatin is then folded to form a densely packed chromosome.

Packing Itself Controls Gene Expression

A consideration of the higher orders of DNA packaging shows that some, if not most, of the bases in the DNA strand must be relatively inaccessible to any cellular enzymes. Think about packing a suitcase to go on a vacation. To avoid wrinkling a special suit or a dress you might make a core out of socks, pajamas, and other items whose appearance is less than critical. Once you've made the core you could carefully wrap it with your special garments, which remain relatively uncreased and wrinkle free. But this type of packing comes with a price: some of your clothes are temporarily inaccessible. If you step in a deep puddle on your way to the airport, you will be unable to get a change of socks from the core without unwrapping the whole setup.

The intricate system of DNA packaging described above also makes some of the DNA in the chromatin fibers inaccessible. Scientists believe that one result of this arrangement of DNA in chromosomes is that some of the base pairs, and therefore some of the genes, are inaccessible to the enzymes necessary for transcription of mRNA. In other words, the lack of transcription of some eukaryotic genes may be a function of the physical structure of the eukaryotic DNA itself.

Evidence for this hypothesis appears in the chromosomes of ***Drosophila***—fruitfly—larvae. *Drosophila* have special arrangements of chromosomes in the tissues of their salivary glands. These chromosomes, known as **polytene chromosomes,** consist of many parallel chromosome strands. Polytene chromosomes, unusual in that they are much larger than a single chromosome, provide an excellent opportunity to observe the physical changes which affect a chromosome as gene expression is regulated.

Microscopic analysis of larval development shows that the structure of the polytene chromosomes varies as development progresses. In fact, **puffs** appear

Factors That Govern Gene Expression 153

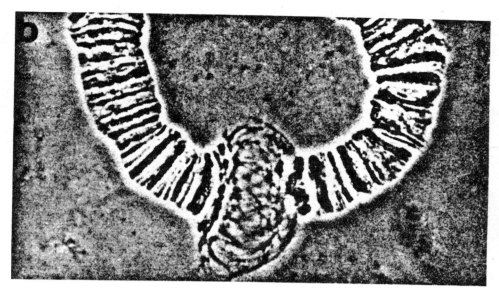

Figure 8.5
A chromosome puff from *chironomus sallidiviltatus*. Chromosome puffs represent specific regions of genetic activity along a chromosome. The location of the puffs changes as development proceeds thus suggesting a relationship between higher order DNA packaging and the control of gene expression.

at specific sites on the chromosome during specific stages of larval development (Figure 8.5). The puffs are predictable in both timing and location on the chromosome. Different puffs appear in different areas as the larvae reach different stages of development. A specific number of chromosomal puffs in predictable locations accompanies each characteristic stage of development. Some observers believe that the puffs represent genes which are being unpacked from the chromosome so that they become accessible to the enzymes of transcription. As different gene products are required for further development, different genes become unpacked, and therefore different puffs become obvious. Thus, the pattern of polytene chromosomal puffs may visually indicate the different genes which play roles in the ongoing process of development.

Puffs Can Be Hormonally Induced

The formation of polytene chromosome puffs can demonstrate the relationship of regulatory molecules to gene expression. The insect hormone **ecdysone** induces the molting process in which the insect sheds its outer covering (Figure 8.6). In addition, scientists have characterized the particular pattern of chromosomal puffs which indicate the onset of the molting process. If ecdysone is applied to an insect which is not naturally ready to molt, the chromosomes take on the puff pattern characteristic of a molting animal. In other words, the hormone itself acts as an inducer for the expression of the genes involved in the process of molting.

Figure 8.6
Molting adult leafhopper, newly emerged from its nymphal skin. Molting is induced by the hormone ecdysone which, in turn, induces a particular pattern of chromosome puffs.

Eukaryotes Have Multiple Gene Control Enzymes

RNA polymerase is the enzyme responsible for the transcription of DNA templates into RNA molecules. A single type of RNA polymerase causes the synthesis of all three types of RNA molecules (Chapter Two) in prokaryotic cells. In eukaryotic cells, however, three different types of RNA polymerase molecules are each responsible for the synthesis of a different type of RNA molecule. Specifically, an enzyme known as **RNA polymerase I** directs the transcription of one of the two types of rRNA molecules. The **RNA polymerase II** enzyme directs the synthesis of mRNA, while the synthesis of a second type of rRNA molecule and of tRNA—the RNA molecule responsible for transporting amino acids to the growing polypeptide on the ribosome—is directed by **RNA polymerase III.** In addition to these three RNA polymerase enzymes, eukaryotic cells also have an enzyme called **poly(A) synthetase** which modifies eukaryotic mRNA molecules by the addition of their characteristic poly(A) tails. The existence of this variety of eukaryotic enzymes, as compared with the single prokaryotic RNA polymerase enzyme, may indicate a wider variety of possible sites for genetic control in the eukaryotic cell. Because a series of different enzymes each carries out its own specific activity, there are more available molecular functions whose modulation affects levels of gene expression.

DNA Sequences which Exert Transcriptional Controlling Effects

The control of both prokaryotic and eukaryotic gene expression is similar in that both rely on the interaction of regulatory proteins with specific nucleotide sequences called **regulatory sites.** We learned that regulatory proteins can be divided into two basic types: **activator proteins** whose presence turns on gene expression and **repressor proteins** whose presence turns off gene expression. In addition, we observed the function of the promoter in the control of prokaryotic operon expression. Promoters have parallels in eukaryotic systems.

Promoters which Interact with RNA Polymerase II

The study of genetic engineering and recombinant DNA technology poses questions regarding the factors that allow the expression of engineered genes after their insertion into appropriate cloning vectors and host cells. Since this expression of exogenous genetic material relies primarily on the transcription of mRNA from the inserted gene, we will concern ourselves primarily with promoters which interact with RNA polymerase II, the polymerase responsible for the transcription of mRNA. For the sake of completeness, it should be noted that the rRNA, tRNA and other molecules which play a role in the transcription and translation of inserted exogenous genes are generally of host origin and are thus under normal host-cell control. The mRNA, however, is transcribed from an exogenous template most often found on a cloning vector and may therefore be under the control of elements included on the cloning vector itself. Thus, we must consider the identity of those controlling elements which may be eukaryotic in origin.

THE TATA BOX In eukaryotic genes, a conserved sequence of base pairs is located approximately 25 base pairs upstream of the transcriptional start site. This sequence, rich in adenine and thymine nucleotides, goes by the name of **TATA box** because of its consensus sequence which reads "TATAAAA." Despite its location somewhat farther upstream, the TATA box appears to be analogous to the Pribnow box of prokaryotes. The TATA box facilitates, but does not appear to be essential to, the process of transcription.

THE CAAT BOX Located some 75 base pairs upstream of the transcriptional start site of many eukaryotic genes is a second conserved sequence of base pairs. This region has a consensus sequence which reads "GG(G or C)CAATCT" and is called the **CAAT box,** or sometimes the **CAT box.** Mutations within the CAAT box decrease the efficiency with which the associated downstream gene is transcribed. Thus, the CAAT box probably directly affects the relative efficiency with which a promoter binds to RNA polymerase II.

Enhancers Promote Transcriptional Activity

Enhancers, sequences of base pairs found in eukaryotes and in some viruses which infect eukaryotic cells, are able to increase the rate of transcription of certain genes. Enhancers differ from promoters in two main characteristics. As described above, promoters retain a fixed spatial relationship with the genes which they govern: promoters almost always contain conserved elements 25 base pairs and 75 base pairs upstream of the target gene. Promoters that are moved from this location and inserted into a new site on the DNA strand rarely retain their full ability to function. In contrast, enhancers can be moved thousands of base pairs away from the target gene and still function efficiently. In addition to this feature, enhancers, unlike promoters, function regardless of their orientation in the DNA strand. That is to say that if an enhancer is removed from the DNA strand, rotated 180°, and re-inserted—now in the reverse orientation—into the strand, it will continue to function by increasing the rate of transcription of the target gene.

Eukaryotic Gene Expression Is also Controlled after Transcription

You recall from our discussion in chapter two that a specific and well-defined pathway must be followed to get from gene to gene product. For any gene to be successfully expressed it must first be (1) transcribed to produce a complementary mRNA molecule. In the case of eukaryotic cells which have both introns and exons in their genes, that mRNA molecule must undergo (2) post-transcriptional modification. The mRNA molecule then becomes the template for the process of (3) translation which produces a polypeptide molecule. That molecule then may be subject to (4) post-translational modification before it takes on the form of a fully functional protein molecule. Although steps one and three are common to both prokaryotes and eukaryotes and have been discussed above, steps two and four are primarily restricted in eukaryotes. Let's look at these steps in a little more detail.

Post-transcriptional regulation refers to the fact that some mRNA molecules must be structurally modified before they are competent to participate in the process of translation. This structural modification generally involves the removal of sequences which are complementary to the non-coding intron regions of the structural genes, the splicing of the remaining mRNA fragments to form one complete molecule, and the addition of the 5' cap structure and the 3' poly(A) tail. Interference with any of these processes affects the ultimate level of gene expression by diminishing or obliterating the effectiveness of the mRNA molecule as a template for translation.

Post-translational regulation refers to the necessity of structurally modifying some polypeptides as they come off of the ribosome/mRNA complex before they acquire their ability to function. This modification may involve, for example, the cleavage of one long non-functional polypeptide into two or more functional molecules, as occurs in the production of insulin. The initial translational product of the insulin gene is a polypeptide known as **proinsulin,** a non-functional insulin precursor composed of 86 amino acids. During post-translational modification the proinsulin molecule is cleaved into two fragments, one of which is 30 amino acids long and the other, 21 amino acids long. These subunits retain their contact with one another via three chemical bridges called **disulfide bridges.** This cleaved molecule is called functional insulin. If the post-translational modification is prevented, the insulin molecule never achieves its fully functional capacity.

Gene expression can also be controlled by variable levels of **degradation** of the gene product. In this situation, degradation refers to the variable rate with which different protein products can be subject to cellular enzymes which cause their destruction. If the rates of transcription and translation remain fairly constant, changes in the rate of degradation alter the amount of protein product available to the cell.

EXPRESSION OF EUKARYOTIC DNA IN PROKARYOTIC HOSTS

This chapter has pointed out that the controlling elements of prokaryotic and eukaryotic genes, while similar, do have significant differences. Therefore, to achieve expression of a cloned eukaryotic gene in a prokaryotic host cell, one must provide the appropriate controlling elements. The most straightforward way to do this is to include the controlling elements on the cloning vector which carries the cloned gene. In this way, the eukaryotic gene is presented to the host prokaryotic cell in a more familiar manner. For example, an ideal expression vector might contain a target gene consisting only of coding exon regions along with appropriately situated prokaryotic promoter sequences and a Shine-Dalgarno site. Expression vectors which allow the expression of eukaryotic genes by prokaryotic host cells have, in fact, been developed along these lines. Some expression vectors also contain the first few codons, including the transcriptional start site, of a bacterial gene—often the beta-galactosidase gene—ligated to the eukaryotic gene insert. The polypeptide produced from such a template is known as a **fusion protein** because it contains both prokaryotic and eukaryotic sequences.

Many possible solutions—all of which may succeed with some level of efficiency—deal with the problem of how to get a cloned gene to be expressed in a host cell. Construction of workable plasmid involves the artistry that is a very real part of scientific research. True finesse in cloning vector construction can come only from a thorough and in-depth understanding of the principles of DNA manipulation and gene control.

REVIEW AND SUMMARY

1. Prokaryotic cells generally live within quickly changing environments. To respond to this situation, prokaryotic genes have developed to the point where they are arranged in groups called **operons.** Operons, are groups of related structural genes along with their controlling sequences, which are transcribed in a coordinated manner.

2. Five basic components make up a bacterial operon: 1) The **regulator gene** encodes a molecule which is sensitive to the presence of certain molecules in the bacterial environment. 2) The **repressor molecule** is encoded by the regulator gene and changes its molecular configuration in response to various environmental conditions. 3) The **operator** is a sequence of bases to which the repressor molecule is able to bind. The operator's location slightly overlaps that of the promoter. 4) The **promoter** is a sequence of bases to which the RNA polymerase enzyme binds prior to the transcription of the associated structural genes. 5) The **structural genes** encode functional enzymes. The operon itself functions to control the level of expression of the structural genes.

3. The *lac* **operon** encodes three enzymes which are involved in the digestion of lactose and is a general model for all repressible bacterial operons. The *lac* operon functions as follows: when the environment surrounding an *E. coli* cell contains lactose, the lactose binds to the repressor molecule. This binding induces a structural change in the repressor which renders the repressor unable to bind to the operator and therefore unable to interfere with the promoter. RNA polymerase binds to the promoter and initiates transcription of the structural genes.

4. Without the presence of lactose in the environment surrounding the *E. coli* cell, the repressor molecule does not undergo any structural changes. The repressor molecule therefore binds to the operator. Because the operator physically overlaps the promoter, this binding of repressor to operator interferes with the ability of the RNA polymerase enzyme to bind to the promoter. Transcription of the structural genes does not take place since the necessary enzyme is not able to bind to the operon.

5. The *lac* operon is an **inducible operon** in that its structural genes are not transcribed except in the presence of an **inducer.** Other operons, for example the ***trp* operon,** are **repressible operons.** Their structural genes are expressed continuously except in the presence of a **corepressor molecule.**

6. **Promoters** are DNA sequences, generally located upstream from the transcriptional start site of a gene, to which RNA polymerase binds, thus allowing transcription to proceed. Various promoters act with varying levels of efficiency.

7. The effectiveness with which a promoter binds RNA polymerase varies with a number of factors, including base sequence and relative distance from the transcriptional start site. Two conserved sequences of bases characterize prokaryotic promoters. These sequences, the **Pribnow box** at 10 base pairs upstream and the **-35 site** at 35 base pairs upstream, have consensus sequences as follows: TATATT and TATTGACA, respectively.

8. **Catabolite activator proteins,** or **CAP** proteins, bind to the **CAP binding site,** located very close to the promoter. The binding of the CAP protein to this site is enhanced if the protein first binds to a molecule of cAMP. CAP proteins are involved in both positive and negative regulation. An example of such positive regulation appears when an *E. coli* cell, deprived of glucose, produces large amounts of cAMP. The cAMP, in turn, enhances the binding of CAP to the CAP binding site of operons which are involved with the digestion of alternative food sources. This results in an increase in the efficiency of RNA polymerase binding, and therefore an increase in the expression of associated structural genes. Negative regulation takes place when the *E. coli* cell is provided with glucose: levels of intracellular cAMP remain static, therefore, the binding of CAP to alternative operons is not enhanced and so does not enhance the binding of RNA polymerase. This chain of events results in the lack of enhancement of the structural genes needed to digest more complex food sources.

9. The **Shine-Dalgarno sequence,** a sequence of base pairs found at the 3' end of an mRNA molecule, is complementary to a segment of one of the rRNA molecules which make up a ribosome. Scientists believe that this sequence ensures correct positioning of the message on the ribosome, so that the correct reading frame is attained in translation. Variations in either the position or the sequence of the Shine-Dalgarno sequence apparently decrease the efficiency of translation. Thus, the Shine-Dalgarno sequence may play a role in gene regulation at the level of translation.

10. The physical packing of DNA molecules may influence control of the expression of eukaryotic genes. This highly complex packing results in the hiding of many sequences of bases from the enzymes in the intracellular environment. Such hiding may play a role in gene control, as suggested by the existence of chromosomal puffs and other observed morphological changes in chromosomes related to gene expression.

11. Eukaryotic genes also have controlling sequences which bear some similarity to controlling prokaryotic gene sequences. These sequences include the **TATA box,** located some 25 base pairs upstream of the transcriptional start site, and the **CAT box,** located approximately 75 base pairs upstream of the start site. Modifications in either of these sequences, or in their positioning, can modify the level of transcription of the involved structural genes.

12. **Enhancer sequences** behave in a fashion similar to promoter sequences, with two main differences. First, these sequences are not dependent on a fixed spatial relationship with their related structural gene for optimum functioning. Second, enhancer sequences function regardless of their orientation within the DNA strand. They can be excised, flipped 180°, and reinserted, and still retain their original function.

Cloning genes in yeast cells

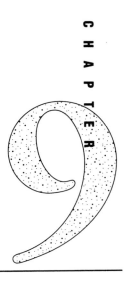

Terry runs Sweet Stuff, a bakery high in the mountains of Colorado. Sweet Stuff specializes in a delicious variety of baked breads and fancy raised pastries. Each morning between one and one-thirty, Terry arrives at Sweet Stuff and begins his morning routine (Figure 9.1). First he mixes the dough for his hearty multi-grain bread. Terry combines 81 pounds of various ground whole grain flours and sprouted seeds with just under one pound each of salt and sugar. In another bowl Terry adds 5 pound of compressed yeast to 37.5 pounds of 110°F water. Terry also adds sugar or honey to the yeast and water mixture. Within a short time the yeast cells begin to grow. Extracting energy from the sweetening agent, the yeast cells divide over and over again. Terry adds the growing yeast culture to the dry ingredients along with a measure of shortening. He places the dough into an industrial sized mixer and switches it on. As the dough hook moves around and around within the mass of dough, gluten develops and the dough becomes firm and takes on a stretchy consistency.

Soon Terry divides the dough and shapes it into loaves. The loaves, dense and firm, are placed in a warm place to rest while the yeast cells do their work. When Terry checks the loaves about ninety minutes later, an amazing transformation has occurred: the yeast cells distributed throughout each loaf have continued to grow and metabolize. As a byproduct of its metabolism, each cell has produced carbon dioxide (CO_2) gas which is released into the dough. The CO_2 cannot escape because of the nature of the gluten in the dough. As a result, many little gluten balloons, each filled with CO_2, form within the bread dough. The presence of those little balloons causes the bread dough to rise and take on a lighter, fluffier texture. Terry, now ready to bake his loaves of multi-grain bread, places the shaped and risen dough in the 375°F oven. He confidently anticipates removing the beautiful, fragrant loaves after about twenty to thirty minutes of baking.

Terry smiles to himself as he remembers his first effort at baking bread at Sweet Stuff. Because he neglected to take the relatively high altitude of 9,000 feet into account, Terry's first loaves were not so perfect. Too much yeast had caused the loaves to rise quickly and collapse, producing a hard, dense product.

Cloning Genes in Yeast Cells

Figure 9.1
Bakers rely on yeast to perform certain tasks for them, so they are considered by some to be a type of biotechnologist.

Terry had experimented, manipulating the quantities of yeast and other ingredients, until he consistently produced a uniformly acceptable and delicious product.

Terry is, in a sense, a biotechnologist. Although he has not recombined DNA molecules or placed excess DNA into his yeast cells, he relies on this relatively simple eukaryotic organism to perform certain tasks for him. When those tasks are not performed to his standards, he manipulates the system by varying the ingredients and conditions until his needs are met. Terry joins a long line of practical biotechnologists. Some of his compatriots use different yeast strains to produces beers and wines, while others rely on bacterial cells to make cheeses.

OVERVIEW This chapter discusses the use of yeast cells in modern genetic engineering. As a relatively simple eukaryotic cell, yeast provides a bridge between prokaryotic host cells and the more complex higher eukaryotic hosts which will be discussed in Chapter Ten. This chapter deals with the physical characteristics of yeast cells and the manipulation of yeast plasmids to produce a variety of cloning vectors. We will discover why yeast cells are not appropriate hosts for the cloning of all eukaryotic genes, examine the characteristics of vectors which lead to the expression of cloned genes in a yeast host cell, and answer the following questions.

1. What are some of the typical vectors used in a yeast cloning system?
2. What is a shuttle vector and why does it shuttle?
3. How does genetic complementation help to identify the various genes of the yeast cell?

YEAST: A SIMPLE EUKARYOTIC HOST

Why Use Yeast as a Host Cell?

We have discussed in some detail the use of prokaryotic cells, especially *E. coli*, as host cells for gene cloning (Chapters Six and Seven). These cells provide remarkable advantages in experimental cloning procedures: they grow easily under laboratory conditions, have a very short generation time, and possess a well-understood genome. Because of these factors, we might consider *E. coli* to be the ideal host in gene-cloning experiments. Unfortunately, that is not always the case, as attempts to clone eukaryotic genes bear out. Eukaryotic genomic DNA is often unsuitable for cloning in prokaryotic hosts due to the presence of both introns and exons in the eukaryotic DNA. In order to clone eukaryotic genomic DNA directly, without first making cDNA constructs, we must rely on eukaryotic host cells which, unlike prokaryotic cells, possess the enzymes required for the removal of genomic introns. In addition, a thorough analysis of the controlling mechanisms of eukaryotic gene expression can only take place in eukaryotic cells themselves. Let us, therefore, examine gene cloning in a relatively simple eukaryotic cell.

Yeast Is a Useful Eukaryotic Host

The yeast **Saccharomyces cerevisiae**—also known as **S. cerevisiae**—is probably the best-characterized eukaryotic organism in existence (Figure 9.2). This yeast, used commercially in bread-making and in the production of beer and other alcoholic beverages, grows easily under laboratory conditions. *S. cerevisiae* has a genome of approximately 2×10^7 base pairs—only four times the size of the *E. coli* genome and 1/150th the size of the human genome. In addition, *S. cerevisiae* populations can double in cell number in only a few hours, quickly providing significant numbers of cells available for research purposes. Along with its ability to synthesize products encoded by genes containing both introns and exons, *S. cerevisiae* has other important advantages over *E. coli* as a potential host cell. For example, mutant strains of *E. coli* tend to revert back to normal phenotype, while mutant yeast strains are far more stable. In addition, while *E. coli* cells don't generally secrete proteins into the growth medium, yeast cells do secrete proteins and it is easier to collect a molecule which has been secreted into the growth medium than one which remains within a cell's interior. All of these characteristics combine to make the experimental manipulation of yeast cells an important part of genetic engineering.

Yeast Cells Contain Both Chromosomes and Plasmids

In addition to seventeen linear chromosomes, cells of the yeast strain *S. cerevisiae* contain a type of plasmid known as the **2 micron circle.** This plasmid, 6,318 deoxyribonucleotide pairs in length, contains two 599 base pair sequences of repeated nucleotides and is present at a level of approximately fifty copies per

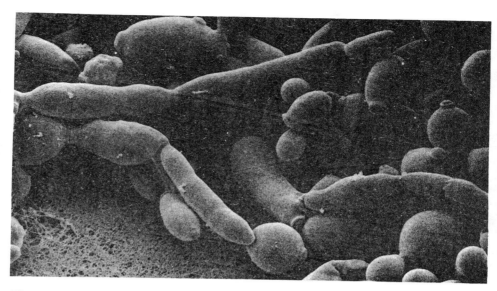

Figure 9.2
The well-known characteristics of the yeast *Saccharomyces cerevisiae* make it an excellent tool in the genetic manipulation of eukaryotic cells. Micrograph of *S. cerevisiae*.

cell. The 2 micron circle contains a single origin of replication along with four genes. Two of those genes help to maintain plasmid expression at the level of about fifty copies per cell while the other two genes appear to be involved in the amplification and recombination processes of the plasmid itself.

Yeast Can Be Converted to Spheroplasts

A wide variety of protocols allow the transformation of yeast cells with exogenous donor DNA. Some of these protocols are directed at the transformation of intact yeast cells and others at the transformation of **spheroplasts** or **protoplasts** (Figure 9.2). These terms refer to the cell construct that remains after the removal of most or all of the yeast cell wall by enzymatic treatment. Although technically a spheroplast retains some portion of the cell wall and a protoplast retains none, the two terms are often used—albeit incorrectly—interchangeably in the context of laboratory research.

Calcium Mediated Transformation of Yeast Spheroplasts

It is relatively easy to transform yeast spheroplasts with exogenous donor DNA by first mixing the donor DNA with a source of excess calcium ions such as calcium chloride ($CaCl_2$) or calcium phosphate ($CaPO_4$). This results in the precipitation of microscopic particles consisting of both DNA and calcium. This precipitate can then be mixed with yeast spheroplasts along with a chemical agent such as **polyethylene glycol,** which not only acts as an antifreeze in your car, but also causes the outer membrane of a cell or spheroplast to become more permeable. The DNA precipitate moves quite easily through the permeabilized membrane of the treated spheroplast, thus entering the cytoplasm of the cell. Transformed spheroplasts can then be plated on a mixture of 3% agar and growth medium. Such a growth environment permits the regeneration of a new cell wall around the newly transformed cell.

PLASMID VECTORS PLAY A ROLE IN TRANSFORMATION OF YEAST

A series of vectors, extremely useful in the various protocols which cause the transformation of yeast cells, has been developed.

YEp: A Yeast-Derived Plasmid

One of these vectors, called **YEp** or **yeast episomal plasmid,** contains the origin of replication from the 2 micron circle. In addition, YEp plasmids often contain a yeast gene which encodes a selectable marker, such as a gene whose product confers antibiotic-resistance to the host cell, along with an origin of replication derived from the *E. coli* chromosome. YEp plasmids, while fairly stable within a transformed host cell, exist with a relatively low copy number and are not integrated into the host cell genome.

YRp: Plasmids that Replicate Independently

Another type of plasmid cloning vector used to transform yeast host cells is the **YRp** or **yeast replicative plasmid.** This plasmid relies on the presence of an **autonomously replicating sequence** for its ability to replicate. This DNA sequence, also called an **ARS**, allows the replication of a plasmid in a host yeast cell to occur independently of the replication of the host cell chromosomes. The ARS sequence may derive from either the yeast strain itself or from some other organism. All ARS sequences appear to contain a consensus sequence of AAA(C or T)ATAAA. The presence of the ARS sequence in a cloning vector increases the frequency of host cell transformation to approximately one in one thousand cells. In contrast, plasmids which do not contain this sequence have a frequency of transformation of approximately one in ten million cells.

In addition to the ARS, the YRp also contains a yeast-selectable marker and an origin of replication of *E. coli* derivation. Although YRp plasmids efficiently transform yeast cells due to the presence of the ARS and occur in relatively high copy numbers within transformed cells, the YRp-transformed yeast cells tend to be highly unstable and revert to an untransformed state as time passes.

YCp: Plasmids with Centromere Fragments

The inclusion on the cloning vector of a 6,000 to 10,000 base-pair DNA fragment, known as **cen,** increases the stability of yeast cell transformation. The cen fragment represents a cloned portion of a eukaryotic **centromere**—the region of a chromosome responsible for the correct movement of chromosomes during the cell division processes of mitosis and meiosis. The presence of cen on a plasmid cloning vector—known as **YCp** or **yeast centromeric plasmid**—appears to direct the movements of the plasmid during mitosis, thus increasing the probability that the plasmid will be correctly apportioned to daughter cells during cell division. The stability of transformation therefore increases as the plasmid is less likely to be lost from the transformed host cell as that cell divides.

YIp: Plasmids that Integrate

Another plasmid cloning vector, useful in the transformation of yeast cells, is the **YIp** or **yeast integrative plasmid.** This plasmid consists simply of a selectable yeast marker gene which has been inserted into a plasmid of *E. coli* origin such as pBR322. YIp has no yeast origin of replication and can therefore be maintained within a transformed yeast cell only by integration into the host genome. Such integration leads to a highly stable transformed cell.

HOMOLOGOUS SEQUENCES DIRECT INTEGRATION

The majority of DNA sequences that become integrated into the host yeast genome do so by a **cross-over event** which takes place between homologous DNA sequences in the plasmid and in the host genome. The homologous cross-over event begins when two homologous DNA sequences come to be positioned so that they lie parallel with one another. Breakage and rejoining of the DNA strands at locations which flank the two homologous sequences cause the exogenous and the endogenous DNA sequences to switch places. This process is similar to the process of crossing over which was described in Chapter Three.

Site-Specific Cleavage and Insertion

One can often manipulate the site of integration of the exogenous DNA by judicious use of restriction endonucleases. When a circularized piece of DNA incorporates into a host yeast cell, it becomes stably integrated into the host genome at the very low level of approximately one in one million cells. Furthermore, the site of integration is fairly unpredictable. If, however, that circular piece of DNA is cleaved with a restriction endonuclease prior to the transformation of the host cell, the situation changes. The linearized plasmid molecule integrates into the host genome with a frequency of one in ten thousand cells—approximately 100 times the frequency of integration of the uncut circular plasmid molecule. Perhaps even more important, linearization of the plasmid allows the site of integration to be directed rather than random: the linearized plasmid often becomes integrated at a site on the host genome which is homologous to the plasmid cleavage site. This directed integration depends on the identity of the restriction endonuclease used to linearize the plasmid because of the homology which often exists between the enzyme recognition sequence and the site of integration. In other words, if a preparation of identical plasmid molecules is divided into two aliquots, and one aliquot is treated with *Eco*RI while the other is treated with *Bam*HI, extremely different results ensue when the linearized plasmid integrates into the same host chromosome. The plasmid treated with *Eco*RI will be cleaved at the sequence GAATTC and the resultant linear molecule will most

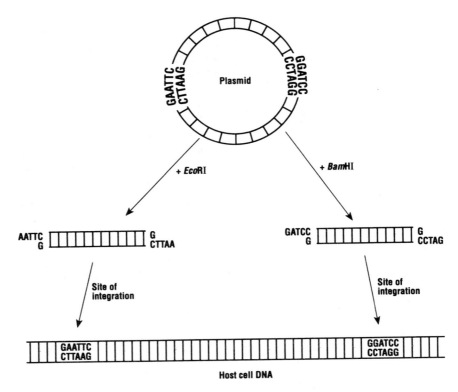

Figure 9.3
Cleavage of exogenous plasmid DNA results in an increase in the level of integration of the plasmid into host cell DNA. Furthermore, the linearized plasmid often integrates into the host DNA at a site homologous to the plasmid cleavage site.

likely integrate into the host chromosome at a site marked by that same sequence of bases. Conversely, the plasmid treated with *Bam*HI will most likely both cleave and integrate at the sequence GGATCC. Thus, variation in the cleavage site can lead to variation in the site of integration (Figure 9.3).

COMPLEMENTATION OF MUTATIONS IDENTIFIES SOME GENES

One of the first ways which scientists developed to identify yeast genes is based on the fact that some yeast genes can **complement mutations** in the *E. coli* cell. In this system, researchers use an *E. coli* strain with a defined mutation— such as a lack of function A due to a lack of enzyme A—as a host cell population. These cells are then transformed with a heterogeneous population of plasmids which was constructed by cleaving the total yeast genome and incorporating the fragments into cloning vectors. The transformed *E. coli* cells are then examined to identify any which have regained function A. The cloned gene in such a cell is then identified as the yeast analog of *E. coli* gene A. While this technique has proven useful, limitations arise in that fewer than half of the yeast genes have a complementary function in *E. coli*. Efficient identification of the remaining non-complementary yeast genes had to wait for the development of a vector which could both be expressed in the yeast cell **and** be propagated in the more easily manipulated *E. coli* cell. Such a vector is called a **shuttle vector.**

YEAST VECTORS CAN ACT AS SHUTTLE VECTORS

Some cloning vectors, including some variations of the examples of yeast plasmid vectors given above, can be constructed so that they can replicate and be expressed in two different host organisms. Such vectors, called **shuttle vectors,** can shuttle an inserted donor gene between two different hosts.

Shuttle vectors prove especially useful in that 1) they can replicate in a eukaryotic host and are therefore potentially useful in the identification and cloning of eukaryotic genes, and 2) they can be transferred from the eukaryotic host—with an inserted gene—into a prokaryotic host which has the tremendous advantage of being easier and faster to grow than a eukaryotic host. Shuttle vectors have been highly useful in identifying yeast genes that were not able to complement *E. coli* mutations.

In such a procedure, total yeast DNA is cleaved and the resulting fragments are inserted into shuttle vectors. Those recombinant molecules are then introduced into *E. coli* cells and allowed to replicate. Once sufficient recombinant plasmid material is available from the transformed *E. coli* population, it is isolated and used to transform yeast spheroplasts with a variety of phenotypic mutations. Any spheroplast that reverts to the non-mutant phenotype as a result of transformation is assumed to have been transformed with a plasmid containing a normal version of the mutant gene. That plasmid—with its newly identified gene—is isolated and re-introduced into *E. coli* cells for easy propagation.

EXPRESSION OF CLONED GENES IN YEAST HOST CELLS

Although scientists initially hoped that the expression of a wide variety of cloned eukaryotic genes that were not easily expressed in prokaryotic host cells would be possible in the yeast host cell, such has not been the case. While some genes are easily expressed others, surprisingly, fail.

Although some yeast genes contain introns, and yeast cells contain the enzymes necessary to process mRNA molecules which contain non-expressed regions, evidence quickly showed that the mechanics of such mRNA splicing are not always the same in yeast as in higher eukaryotes. Experiments suggest that the removal of introns in yeast relies on the presence of a particular sequence of eight base pairs which marks the location of the intron to be removed. This specific eight base-pair sequence does not appear in the genes of higher eukaryotes whose introns are, therefore, often not correctly processed by enzymes of the yeast cell. As a result, eukaryotic genes are often more easily cloned from a cDNA transcript despite the use of a eukaryotic yeast host cell.

Another possible explanation for the difficulty of expressing cloned genes in a yeast host cell may involve the genetic code. Although the genetic code is the same for all organisms, variations in its usage occur among different species. As described in Chapter Two, the genetic code is degenerate, that is, although a single codon specifies one, and only one, amino acid, some amino acids are specified by more than one codon. An analysis of the codons which are used most often in different species yields surprising results: some species use particular codons more frequently than other species. For example, while species A nearly always uses the codon UUU to encode the amino acid phenylalanine, species B uses the codon UUC most often to encode that amino acid. If a gene from species B is to be cloned in host cells of species A, we may encounter a problem: since species A does not contain many copies of the codon UUU, it may not contain many tRNA molecules which are specific for that codon and may therefore be unable to successfully translate the genetic message from species B. Thus, the number of tRNA molecules for a specific codon may be a limiting factor in the expression of higher eukaryotic genes in yeast cells.

Expression Vectors May Aid Transcription in Yeast

As described in Chapter Eight, expression vectors contain, among other things, a promoter sequence which, in turn, facilitates the transcription of the inserted gene. Such a situation can lead to the production of high levels of target mRNA molecules and, therefore, of the inserted gene's protein product. This technique, attaching a promoter to a gene which is to be cloned in a eukaryotic host, was first used in 1981: the promoter of the yeast gene for the enzyme **alcohol dehydrogenase** was inserted into a cloning vector along with the human gene which encodes a type of protein which is active in immunological responses and is known as **interferon**. When this vector was incorporated into host yeast cells, high levels of expression of the human gene were observed. Researchers later determined that expression of the human gene was being driven by the activity of the yeast promoter sequence. Similar vector manipulations, also useful in other types of host cells, will be discussed in Chapter Ten.

REVIEW AND SUMMARY

1. Yeast cells work well as eukaryotic host cells due to the relative simplicity of their genome which, while being 4 times the size of the *E. coli* genome, is only $1/150^{th}$ the size of the human genome. This small size has allowed extensive analysis of the yeast genome. In addition, yeast cells can be grown easily under laboratory conditions.

2. The conversion of yeast cells to **spheroplasts** or **protoplasts** requires the enzymatic removal of most or all of the cell wall. Spheroplasts can be transformed by exposing them to the precipitate which results when DNA molecules are mixed with a source of excess calcium ions. Under appropriate growth conditions, transformed spheroplasts can regenerate a cell wall.

3. The yeast plasmid, called the **2 micron circle,** consists of 6,318 base pairs. The 2 micron circle contains two 599 base-pair sequences of repeated nucleotides and occurs at a level of approximately fifty copies per cell. This plasmid contains an origin of replication along with four genes, two of which are involved in maintaining the level of plasmid in the cell, and two of which are involved in the amplification and recombination of the plasmid itself.

4. A series of vectors has been developed which can be used in yeast-cell transformation. **YEp** contains the 2 micron circle origin of replication and, often, a yeast-selectable marker. YEp has a low copy number and is fairly stable within a transformed cell. **YRp** contains an **ARS** which allows

independent replication of the plasmid. In addition, YRp also contains a yeast-selectable marker and an *E. coli* origin of replication. YRp has a high copy number and is highly unstable within a transformed cell. **YCp** contains a cloned portion of centromeric DNA which greatly increases its stability within the transformed cell. **YIp** contains a yeast-selectable marker in an *E. coli* plasmid. Because it lacks an origin of replication, YIp must integrate into the host chromosome to be maintained. Once integrated, YIp is extremely stable.

5. A linearized plasmid can integrate into a host chromosome more easily than a circular plasmid. In addition, the linear plasmid more likely integrates into a site homologous to the end of the plasmid molecule. Therefore, linearization of the circular plasmid by cleavage at different locations within that plasmid often leads to integration at different sites within the host chromosome.

6. Yeast genes are often identified by their ability to complement mutations in bacterial cells. If an unknown yeast gene transforms a bacterial cell with a known mutation, and if that mutation reverts to wild-type as a result of transformation, it can be deduced that the yeast gene on the transforming plasmid is the equivalent of the mutant bacterial gene. Such a system is called **genetic complementation.**

7. The development of **shuttle vectors,** which have both yeast and bacterial origins of replication and can therefore be maintained in both cell types, has allowed the identification of genes within the yeast cell itself. In this system, a yeast DNA library is made and propagated in *E. coli* cells. When sufficient plasmid DNA is available, those plasmids are then isolated from the *E. coli* cells and introduced into yeast cells with known mutations. Any cloned insert genes which result in genetic complementation of the yeast mutation can be identified. Identified plasmids are then reintroduced into *E. coli* cells for ease of growth.

8. Although many eukaryotic genes can be expressed in yeast host cells, this system is not appropriate for all higher eukaryotic genes. The apparent lack of an appropriate consensus sequence to mark introns for removal appears to prevent the appropriate expression of many eukaryotic genomic inserted genes. In addition, differential use of particular codons by cells of different species may result in an effective lack of specific types of tRNA molecules. This lack may cause a bottleneck which prevents efficient expression of exogenous genes. When expression is possible it is optimized by the use of an expression vector which contains a promoter sequence physically linked to the inserted gene.

Cloning genes in mammalian hosts

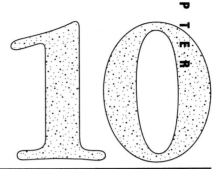

Imagine a room filled with mice. Cage after cage made of gleaming stainless steel or shiny see-through polycarbonate plastic fills countless shelves arranged throughout the room (Figure 10.1). Each individual cage houses from six to ten small white laboratory mice and contains a food supply and a freshly filled water bottle. Each cage has its own assigned place somewhere on the hundreds of feet of shelving in the room. In front of each cage a small white tag contains words of identification. One tag reads:

>Tissue Plasminogen Activator
>For use in cases of cardiac blockage.

Another tag reads:

>Human Growth Hormone
>For use in cases of congenital short stature
>due to lack of endogenous hGH.

And another reads:

>Cachectin Inhibiting Protein
>For use in combatting the wasting
>process commonly seen in cancer patients.

What if our imaginary mouse-filled room became the drug factory of the future? Instead of being filled with stainless steel machines and countless test tubes, this factory would be populated with mice. Each mouse in the imaginary factory contains a special target gene in some of its cells. Each target gene encodes a different pharmaceutical product; the gene which encodes human growth hormone actually synthesizes that molecule and deposits it in the fluid which bathes the cells of one of the mice. Workers at the drug factory collect tissue

Figure 10.1
Will mice be the drug factories of the future?

fluid from that mouse and extract the growth hormone, which can then be purified, packaged, and sold. The mouse then returns to its cage to go on manufacturing more product for the drug factory. When the mouse mates and produces offspring, the members of that new generation are born with the human growth hormone gene just like their parents. Each different cage of mice can be busily manufacturing a different product for our factory of the future.

Does the scene sound far-fetched? Well, it is, but not in its basic idea. Let's see why.

OVERVIEW This chapter discusses the use of mammalian cells as hosts for the cloning of donor genes. It compares the techniques which lead to transformation of mammalian host cells with those which are used in the transformation of prokaryotic bacterial host cells. In addition, it covers some of the methods that allow detection of mammalian host cells which have been transformed with exogenous DNA, as well as the role such transformed cells have played in furthering our understanding of the mechanisms which control the expression of eukaryotic genes. This chapter also address as the following issues:

1. What, exactly, is a human/mouse hybrid and how does such a thing come about?

2. How can viruses act as helpers?

3. How can a high-speed printing press help us sort cells?

4. Why are bacterial enzymes sometimes important to mammalian cells?

MAMMALIAN CELLS AS HOSTS

Why Clone Using Mammalian Cells?

As more and more experiments involving the cloning of eukaryotic genes in both prokaryotic and yeast host cells have been completed, a number of instances have occurred in which expression of the inserted eukaryotic gene simply does not occur in either of these hosts. Despite the presence of the correct controlling sequences which govern gene expression at the levels of both transcription and translation, reasonable levels of expression of some inserted eukaryotic genes have been impossible to attain in either *E. coli* or yeast cells. Genes which fall into this situation can only be cloned successfully in mammalian host cells.

Mammalian cells have additional benefits in their role as host cells: the use of mammalian hosts allows scientists to examine and investigate the regulation and expression of mammalian genes. Such questions can only be thoroughly explored when the target gene is in place in a mammalian host-cell system.

In addition, significant marketing reasons exist for cloning genes within mammalian hosts. Imagine that you have successfully cloned a gene which encodes a valuable human hormone, and now have a large population of *E. coli* cells which churn out large batches of this hormone around the clock. Upon your decision to make your fortune selling this hormone, you isolate your hormone product and put it on the market. Unfortunately, the recipients of this product soon experience negative side effects—generalized low-grade fever and muscle aches and pains—traced back to the presence of **endotoxins** in your hormone preparation. Endotoxins, components of the outer membranes of certain bacteria, including *E. coli,* cause fever and achiness in humans. Eventually, you are forced to remove your product from the market. To bring it back, you must remove all of the endotoxin contaminant coming from the *E. coli* host cells in which you are cloning the hormone gene. Unfortunately, this complicated chemical process would eat into your profits, rendering your whole scheme untenable. You therefore decide to stop using *E. coli* as a host and switch instead to a mammalian host, since a mammalian host cannot contaminate your product with bacterial endotoxins. Unfortunately, in real life nothing is this easy, and although mammalian host cells have attractive features they also have their drawbacks. Both sides of the story will be discussed later in this chapter.

DNA INTO MAMMALIAN HOSTS

Somatic Cell Hybridization

Scientists have long asked questions regarding the effect of placing exogenous DNA into mammalian cells. Until the mid-1970s, however, the only way to address this issue was through the technique of **somatic cell hybridization.** In this technique somatic cells from two or more species are caused to join together or **fuse** with one another. As a result of this fusion, which is induced by a **fusogen,** an agent which induces fusion, such as **polyethylene glycol** or **Sendai virus,** the outer membranes of two so-called parent cells melt together at their point of contact. This results in the formation of a single large membrane, generated in part from one parent cell and in part from the other, which encloses cytoplasm and nuclei from each of the parent cells. With time, the two nuclei also come together and fuse, resulting in the mixing of the chromosomes originally derived from the two different parent cells. Such hybrid somatic cells can be generated from two cells of almost any pair of species as long as the parent cells can be grown in **tissue culture medium**—growth medium for mammalian cells—in the laboratory. This appears to be the simplest way of introducing foreign DNA into a mammalian host cell. The technique does, however, have its drawbacks.

When hybrid somatic cells are first produced they generally contain the full complement of chromosomes from each of the parental cells. However, as time progresses, the chromosomes of one of the parents are randomly lost, or **segregated,** from the cell. Eventually, as a result of this random segregation, most hybrid cells end up back in the original non-hybrid state. That is, they contain the chromosomes of only one parent, all other chromosomes having been segregated. Interestingly, for any given pair of parental species, the chromosomes of one particular species always preferentially segregate. For example, hybrids constructed of mouse and human cells always segregate the human chromosomes while retaining the mouse chromosomes, while hybrids of mouse and rat cells will preferentially segregate the mouse chromosomes and retain the rat chromosomes.

The chromosomal segregation observed in hybrid somatic cells occurs at random. That is, while one particular hybrid cell segregates all human chromosomes except #3, #8, and #11, another hybrid cell segregates all but #4, #16, #22, and X. This random segregation can be modified with the use of a **selective medium.** Such growth media contain one or more chemicals which interact with the products of genes on particular chromosomes. If those gene products are not present, indicating that those particular chromosomes are not present either, the cell will be unable to survive in the selective medium. This means that cells which can live in the selective medium have retained a particular chromosome. Different selective media exist for a variety of human chromosomes, but the segregation of most chromosomes in any one hybrid cell remains random.

Although somatic cell hybridization has enabled scientists to examine the gene products of a great many different genomes and has contributed greatly to attempts to identify the location of genes within the human genome, it also presents problems: 1) hybrid somatic cells are not stable—over time they tend to segregate the chromosomes of one of the parents. 2) The chromosomal complement of hybrid somatic cells is unpredictable—chromosomal segregation occurs randomly. 3) Studies involving hybrid somatic cells deal primarily with entire chromosomes or large chromosomal fragments. When we consider that each human chromosome contains thousands of genes, we can see that the resolution offered by this system is not sufficiently fine. These and other reasons led scientists to search for ways to insert small amounts of foreign DNA into mammalian cells.

Calcium Aids DNA Uptake

Although it had been known for some time that tumor virus DNA, previously stripped of its viral coat, had the ability to enter and transform normal mammalian cells, it was not until 1973 that scientists began to understand how to cause naked DNA to enter a host cell. At this time scientists F. Graham and A. J. Van Der Eb and colleagues at the University of Leiden in the Netherlands developed a novel technique for the transformation of normal mammalian cells using naked viral DNA fragments. These researchers first mixed the DNA fragments with calcium phosphate and then applied the resulting co-precipitate of DNA and calcium ions to mammalian cells which were growing in a **monolayer,** or single layer, under laboratory conditions. When they compared the number of transformed cells resulting from this procedure with the number that resulted when DNA was applied to the potential host cells in the absence of calcium, they found that the presence of calcium greatly increased the efficiency of host cell transformation.

Although the mechanism with which the calcium/DNA precipitate causes transformation of host cells remains uncertain, researchers think that host cells ingest the precipitated material through **endocytosis**—a process whereby the cell membrane folds itself so that it surrounds particles, usually of food but sometimes of DNA and calcium, and then pinches off from the membrane toward the interior of the cell so that a pocket forms and carries the particles into the cell's cytoplasm (Figure 10.2). Some sources suggest that a small percentage of the endocytized DNA actually incorporates into the host cell's genome. This whole procedure is extremely inefficient—most cells never endocytize any DNA at all, and of those that do, most fail to incorporate the DNA into the genome. As a result, the relatively few cells which do successfully become transformed are almost hidden against a background of untransformed cells.

Microinjection

One way to get around the relative inefficiency of the technique of calcium-mediated DNA uptake would be to put a tiny needle on a tiny syringe, fill it with a solution containing cloning vectors with their inserted genes, and simply inject the exogenous DNA into a chosen host cell. Such a technique actually exists and goes by the name of **microinjection.** In this technique a glass micropipette is placed in a microforge and a portion of it is heated until the glass becomes somewhat liquified. The micropipette, then quickly stretched, forms a very fine tip at the heated area. The tip, generally about 0.5 microns in diameter, resembles an injection needle in its hollow design. When cool, the microneedle is placed in a microinjection apparatus which allows the experimenter to control its movements with great precision while watching its progress through the oculars of a

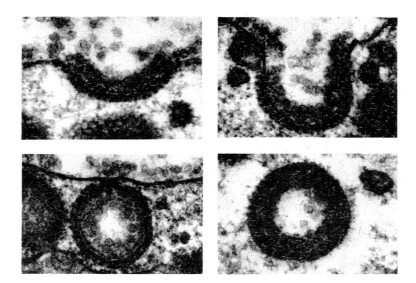

Figure 10.2
Endocytosis involves the formation of a pit in the cell membrane. As the pit gradually deepens, it encloses extracellular materials—food particles, DNA, or a variety of other substances—and finally brings them to the interior of the cell.

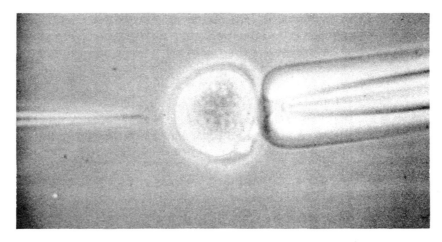

Figure 10.3
Microinjection allows the insertion of exogenous DNA molecules directly into a host cell.

powerful microscope. At the same time a holding pipette, placed in the field of view of the microscope, holds a target cell gently against its tip by the application of a small amount of suction in much the same way that you could hold a piece of paper against a straw by inhaling through that straw. The microneedle can then be guided to the target cell and its tip inserted or injected through the membrane of the cell. Once inside the cell, the contents of the needle can be forced into the cytoplasm, and the empty needle withdrawn. (Figure 10.3).

The technique of microinjection has drawbacks in that it requires complex specialized equipment and skilled operators to manipulate that equipment. On the other hand, microinjection wastes less DNA: injection of a DNA sample into a cell fairly assures its fate—as much as fifty percent of the DNA stably integrates into the host cell genome. If, however, you precipitate DNA with calcium and apply it externally to cells, the vast majority of that DNA will be lost or degraded by the potential host cells. To counteract this factor it should be noted that a multitude of cells can be treated within a given time frame with the calcium-mediated technique, while a good microinjectionist may be limited to ten or fifteen cells per minute. Perhaps the primary advantage of the microinjection technique is that it does not require a selectable marker to be carried along with a non-selectable target gene to aid in detection of transformed cells. Because of the high efficiency of microinjection, it often becomes possible to detect the inserted genes and—because of the high level of stable integration—maintain them in a population without the application of selective media.

Microinjection of Genes into Fertilized Mouse Eggs

One of the most interesting types of mammalian cells used as a host for cloned exogenous genes is the fertilized mouse egg. In this system, purified DNA is microinjected into one of the **pronuclei**—the nuclei of the formerly separate egg and sperm—of fertilized mouse eggs or mouse **zygotes**—embryos at the earliest stages of development. The treated fertilized eggs or zygotes are then implanted into the uterus of a mouse and carried to full-term development.

This type of experiment helps scientists examine complex questions of gene control. How do the descendants of an exogenous gene, inserted early in development, differ from one another in the differentiated cells of the fully developed organism? Will the product of a gene have similar effects on the tissues of different organisms? Will such a microinjected gene be passed on to the offspring of the treated organism? Does the site of integration of the exogenous DNA affect patterns of control and expression? These are just a few of the many questions that can be asked.

The first of this type of experiment was completed in the early 1970s when a group of researchers microinjected DNA isolated from the SV40 virus into mouse embryos at approximately three days after fertilization. These embryos, called **blastocysts,** consist of a hollow ball of cells arranged in a single layer and have about 18 days of further maturation to complete prior to birth. About forty percent of the live births resulting from the reimplantation of these microinjected embryos into host mothers contained evidence of SV40 DNA. This foreign DNA was found in a **mosaic pattern** throughout the mouse tissues. That means that not all organs in a single animal showed evidence of the SV40 DNA, and of those organs containing the viral DNA, not all of the cells within that organ had been transformed. Furthermore, different animals tested positive for SV40 DNA in different organs or tissues. This research suggested that 1) foreign DNA can be incorporated stably into the chromosomes of the developing host embryo, 2) that incorporated DNA can be replicated along with the endogenous cellular DNA and passed along to offspring cells, and 3) that the site of DNA incorporation

Figure 10.4
These littermate mice differ only by the presence of the gene for rat growth hormone in the larger animal. This transgenic animal was produced by microinjecting multiple copies of the rat growth hormone gene into one of the pronuclei of a fertilized mouse egg which was then implanted into the uterus of a host mouse and allowed to grow to term.

appears to be random. Evidence of microinjected DNA found in both somatic and germ cells suggested that in some cases this DNA could be passed on to offspring along with the normal complement of mouse cellular DNA. Later experiments involving the breeding of mice which had developed from microinjected eggs or early embryos confirmed this suggestion. Further experiments also examined the feasibility of microinjecting mammalian DNA, rather than viral DNA, into mammalian organisms.

Microinjected Mammalian Genes Are Expressed in Mice

Figure 10.4 shows two highly unusual mice. These mice have many similarities—they are both grey, have black eyes, whiskers and generally look quite similar to one another but for one enormous difference. Specifically, one mouse grew extremely large while the other is a normal-sized mouse. How did these two mice—littermates who should bear a great resemblance to one another—come to vary so greatly in size? In 1983, scientists Dr. Ralph Brinster and Dr. Richard Palmiter at the Universities of Pennsylvania and Washington, respectively, isolated the gene which encodes the rat growth hormone molecule and linked it to the promoter associated with the metallothionein gene—a gene which encodes a metal-binding protein. Using the techniques of microinjection, they inserted this engineered DNA segment into early mouse embryos to create a **transgenic** animal—an animal which contains foreign DNA stably incorporated into its genome. The engineered embryos were then implanted into the uterus of a host-mouse mother. Would the newborn mice retain the inserted mammalian growth hormone gene? If so, would that gene be expressed and at what levels? If the gene was expressed, would the presence of non-mouse growth hormone in the developing mice have any effect on the phenotype of the animals?

Examination of the live-born littermates showed that some of the mice had indeed retained the inserted growth hormone gene and appeared to have incorporated it into the endogenous genome. Furthermore, those incorporated genes appeared to function under the direction of the metallothionein promoter by producing rat growth hormone in response to induction signals which affected the promoter itself. The presence of the rat growth hormone caused some of the mice to attain a size not normally seen in non-engineered animals: some of the mice were twice as big as their littermates! When examined, the large mice showed much higher rat hormone levels than the endogenous mouse hormone levels of their normally sized siblings. Similar microinjection procedures have also been carried out in other mammalian hosts including sheep and pigs.

Unfortunately, mammalian organisms such as mice, sheep, and pigs cannot currently fill the need for large-scale production of gene products used in human therapeutic settings due to the possibility of viral and bacterial contamination: microbes able to live in a human being and cause disease can also live in a mammalian host cell, thus raising the possibility of transferring infection from the host organism to the human patient right along with the therapeutic molecule. For this reason most large-scale cloning of eukaryotic genes—especially those whose products are intended to have therapeutic uses in humans—continues to be done in *E. coli* or other bacterial hosts; although bacterial contamination may result, human pathogens do not survive in bacterial cells and thus cannot be transferred from cloning host to human patient.

SELECTING FOR TRANSFORMED HOST CELLS

As discussed in Chapter Seven, analyzing a population of bacterial cells in order to detect a subpopulation composed of a relatively few transformed cells involves the presence of a marker. The identity of the particular marker determines the actual selection methods; for example, a gene product marker suggests using a selective medium, and if a marker is a particular sequence of base pairs we may choose a radioactively-labelled nucleic acid probe.

TK as a Marker

Perhaps the best-characterized marker used in selection processes involving mammalian cell populations is the gene which encodes the enzyme **thymidine kinase** or **TK.** This enzyme functions in mammalian cells to salvage thymidine compounds which result from the degradation of DNA and recycles them by reincorporating them into newly synthesized molecules of DNA. Normal functioning of a cell does not require the presence of this enzyme because mammalian cells can also synthesize new thymidine molecules by an alternative pathway. Cells which lack the TK enzyme, referred to as **TK$^-$**, can be isolated by growing a population of cells in a growth medium which has been supplemented with the chemical **bromodeoxyuridine (BrdU).** BrdU, very similar in structure to thymidine—it is a **thymidine analog**—can be incorporated into newly synthesized molecules of DNA in the presence of the TK enzyme. Such incorporation of BrdU however, kills the cell. Thus, the only cells able to live in the presence of BrdU are those which lack the TK enzyme and are therefore unable to incorporate BrdU into DNA molecules.

The TK⁻ cells which are isolated in this fashion rely on their ability to produce newly synthesized thymidine from an alternative precursor rather than by salvaging thymidine from the degradation of DNA. If, however, TK⁻ cells are placed in a selective medium called **HAT medium**—so named because it contains the chemicals **H**ypoxanthine, **A**minopterin, and **T**hymidine—they will not grow. This results because hypoxanthine and aminopterin prevent the synthesis of new molecules of thymidine, while the thymidine already present in the HAT medium cannot be salvaged due to the lack of the TK enzyme. Without a supply of thymidine to take part in DNA replication, cells cannot grow and divide.

If, however, TK⁻ cells incorporate a functioning TK gene into their genomes, they will survive in the HAT medium. Thus, the TK gene itself, in combination with HAT medium, is a useful marker of mammalian cell transformation. In 1977 Dr. Michael Wigler, then at Columbia University in New York, and a group of his colleagues carried out a series of experiments in which they exposed TK⁻ mouse cells to calcium precipitates of the TK gene isolated from *Herpes simplex* virus particles (HSV-TK). In this way they discovered that TK⁺ mouse cells, resistant to the lethal effects of HAT medium, could be generated by mixing the TK⁻ cells with HSV-TK DNA and calcium: The TK⁻ host cells were being transformed with the calcium-precipitated HSV-TK DNA.

Coprecipitation of TK and Non-Selectable DNA

Although the TK gene can be detected with relative ease in a host cell, many genes are difficult to detect. For this reason scientists wondered if the TK marker gene could combine with such **non-selectable** genes and act as a marker for these unmarked genes. Experiments showed that mixing the TK gene with non-selectable DNA, co-precipitating the mixed DNA with calcium, and applying the calcium/DNA precipitate to potential host cells resulted in the transformation of some host cells with a mixture of the non-selectable DNA and the selectable TK marker DNA. Examination of cells treated only with precipitated non-selectable DNA failed to show any significant level of transformation. This suggested that the TK marker could be used to isolate cells capable of taking up exogenous DNA. Scientists hypothesized that such cells took up DNA regardless of the specific type of DNA. The presence of the selectable marker served primarily to identify the newly transformed cells. The marker DNA itself did not appear to cause or enhance the transformation of host cells in any way. Although transformation probably also occurred in experiments which involved non-selectable DNA alone, such events were difficult, if not impossible, to identify.

The Alu Sequence as a Human Marker

Other genetic markers exist in addition to the function of the thymidine kinase gene. One of the most useful is the human DNA marker known as **Alu.** As was described in Chapter Eight, the Alu sequence, a repetitive sequence of base pairs, contains the recognition site for the restriction endonuclease *Alu*I. This sequence, repeated between 100,000 and 300,000 times within the human genome, is unique as a highly repeated sequence to the human genome. The presence of the Alu sequence can therefore be used to distinguish human DNA sequences from DNA sequences of other species. In other words, the Alu sequence acts as a marker for human DNA. Thus, by examining potential host cells with a radioactively labelled probe specific for the Alu sequence, scientists can detect those host cells which have been transformed by the incorporation of human DNA sequences.

Expression of Donor Genes Can Be Transient or Stable

The successful detection of transformed host cells does not necessarily predict the fate of the exogenous DNA which was incorporated. The DNA fragments taken up by mammalian cells appear to become linked to one another sometime during the process of transformation. As a result of this linkage, the collection of DNA fragments which enters a single host cell converts into a single polymeric molecule consisting of multiple units of exogenous DNA linked end to end. One of three possible fates then await this molecule. In the first possibility the molecule immediately degrades and its components are either recycled or excreted from the cell, with no measurable expression of any of the genes contained on the exogenous DNA molecule. The second option allows the molecule to remain within the cytoplasm of the cell; the genes on this molecule may be expressed for a limited time. Eventually the molecule goes on to be degraded and lost from the cell, thus accounting for the limited or **transient** nature of the gene expression. The third fate incorporates the polymeric molecule into the host cell genome. If this occurs, gene expression may be quite **stable** as the exogenous DNA replicates and passes on to daughter cells along with the chromosomes of the host genome.

CLONING VECTORS AND EUKARYOTIC HOSTS

As discussed in Chapter Eight, the elements which control the expression of genes in eukaryotic systems differ from those which control gene expression in prokaryotic systems. For this reason cloning vectors that function in eukaryotic cells and lead to the expression of the target gene fragments obviously differ somewhat from prokaryotic vector systems. A eukaryotic cloning vector must contain a eukaryotic origin of replication, eukaryotic selectable markers, controlling elements appropriate to eukaryotic transcription and translation, and the appropriate signals for post-transcriptional modification of mRNA molecules. Examples of useful mammalian cloning vectors have been derived from SV40 virus particles. Let's examine two of these engineered vector systems more closely.

VIRAL VECTORS AND MAMMALIAN HOST CELLS

Although transformation of host mammalian cells is possible using the cloned TK gene in a coprecipitation protocol, such procedures require the use of a mutant TK$^-$ host cell: transformation events would be virtually undetectable without the TK$^-$ mutation. However, in some cases the only possible host, or the best host, is **wild-type**—normal and non-mutant. In this case one must use a cloning vector which contains a marker gene that can be detected against the background of a wild-type cell. Examples of such cloning vectors have been developed by modifying the genome of a virus known as **SV40.**

SV40, or **Simian Virus 40,** a small tumor virus, contains a single double-stranded circular chromosome and normally infects monkey cells. SV40, like bacteriophage lambda (Chapter Six), can establish both lytic and lysogenic infections in its host cells. In the lytic infection, multiple copies of the infecting virus particle are produced inside the host cell. Eventually the host cell lyses and newly synthesized virus particles spread throughout the immediate environment. In the lysogenic infection, infected cells become transformed by the incorporation of the viral chromosome into the host genome. Such cells take on a tumorous phenotype and are able to induce tumor formation even when they are injected into other host animals.

The genes of the SV40 chromosome can be divided into two sets which are distinguished by the timing of their expression: the **early genes** encode viral tumor antigens, and the **late genes** encode proteins which make up the viral coat of the SV40 particle itself. In addition to these early and late genes, three short lengths of DNA which serve no apparent essential function have also been identified. Scientists have been able to replace the non-essential portions of the SV40 viral genome with DNA which is to be cloned in much the same way as bacteriophage lambda functions as a cloning vehicle. The SV40 particles containing an engineered chromosome then go on to infect host cells, thus resulting in the replication of the inserted gene as the infection proceeds.

Unfortunately, the relatively small, non-essential and easily replaced portions of the SV40 genome are not large enough to accommodate all inserts. Therefore, sometimes researchers must replace essential portions of the viral genome with the DNA which is to be cloned. When this occurs, the SV40 particle can no longer carry out efficient infection of the host cell and is therefore no longer an efficient cloning vector. To remedy this situation, the defective engineered viral particles are mixed with normal, non-defective, **helper** virus particles. In this way the helper particles provide the genetic products which may be missing from the defective particle. The presence of helper viral proteins allows the defective virus to proceed normally with infection, thus resulting in the cloning of the inserted lengths of DNA.

SV40 Vectors Can Carry Selectable Markers

As we have seen, the transfection of mammalian host cells, a highly inefficient process, can be aided by the presence of a selectable marker, such as the TK gene, or by the ability to transfect a cell with an engineered viral cloning vector. Scientists have combined these factors to produce two systems in which an SV40 viral cloning vector also contains a selectable marker.

The first of these systems depends on the metabolic pathway which mammalian cells use to generate **purine nucleotides**—purine bases, such as adenine and guanine, which are chemically bound to either a deoxyribose sugar or a ribose sugar and to one or more phosphate groups. In wild-type mammalian cells this synthesis reaction relies on an enzyme known as **hypoxanthine guanine phosphoribosyl transferase** or **HGPRT**. This enzyme functions in two separate biochemical reactions. First, HGPRT aids in the conversion of the

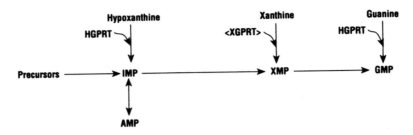

Figure 10.5
As described more fully in the text, mammalian cells rely on the enzyme HGPRT for the conversion of various precursors into the purine nucleotides, including guanine monophosphate or GMP, necessary for cell growth. By inactivating the HGPRT enzyme with mycophenolic acid, scientists can select for cells which have been transformed with the bacterial GPT gene, because its product—the XGPRT enzyme—replaces some of the lost HGPRT function, thus allowing transformed cells to survive in selective media.

base **hypoxanthine** to **guanosine monophospate** or **GMP,** a guanine nucleotide. This conversion takes place through a series of two precursors known as **IMP** and **XMP.** In addition, HGPRT can convert guanine directly to GMP (Figure 10.5). The enzyme HGPRT is also able to aid in the conversion of xanthine to XMP, albeit at a very low level of efficiency. Addition of the chemical **mycophenolic acid** to the growth medium blocks the primary action of the HGPRT enzyme. A cell living in such an environment must rely on the inefficient conversion of xanthine as the sole means of synthesizing the needed GMP. Thus, a wild-type mammalian cell soon expires in a growth medium supplemented with mycophenolic acid.

Bacterial cells possess a gene which encodes an enzyme closely resembling the mammalian HGPRT. The bacterial enzyme, known as **xanthine guanine phosphoribosyl transferase** or **XGPRT,** is encoded by the **gpt** gene. XGPRT functions efficiently in the conversion of xanthine to XMP which can in turn be converted to GMP. To exploit this fact, scientists have constructed a cloning vector in which the *E. coli* gpt gene is inserted into an SV40-based vector so that the SV40 promoter sequence controls XGPRT synthesis. Mammalian cells can be transformed using this **SV40gpt** vector. Such transformed cells express functional XGPRT enzyme, thus enabling them to utilize xanthine as a source of the GMP necessary for continued growth.

The ability of mammalian cells to produce functional enzyme molecules from the bacterial gpt gene has allowed the development of a selection system based on chemical manipulation of the growth medium which is used. In this system, wild-type mammalian cells, subjected to transformation procedures involving the SV40gpt cloning vector, are then placed in growth medium which has been supplemented with mycophenolic acid and xanthine. Cells which fail to take up the SV40gpt vector cannot survive in this medium because of their inability to synthesize the required GMP due to the inactivation of the HGPRT enzyme by mycophenolic acid. On the other hand, mammalian cells that have successfully incorporated the SV40gpt vector contain XGPRT as a result of transcription and translation of the bacterial gpt gene contained on the vector. The XGPRT enzyme

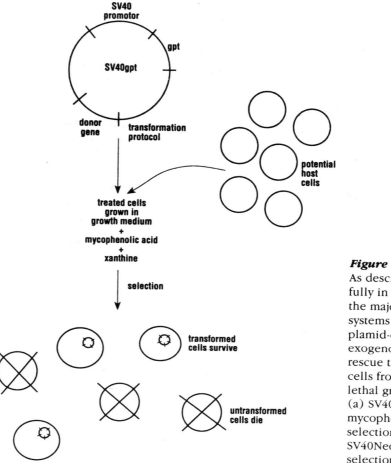

Figure 10.6
As described more fully in the text, two of the major selection systems make use of plamid-carried exogenous genes to rescue transformed cells from otherwise lethal growth media. (a) SV40gpt/ mycophenolic acid selection system. (b) SV40Neo/G418 selection system.

enables the transformed cells to convert xanthine to XMP and from there to GMP. Thus, the simple ability of a wild-type host cell to grow and divide in a medium containing mycophenolic acid and xanthine indicates successful transformation of the cell with a vector containing the gpt gene along with the donor gene which is to be cloned (Figure 10.6).

The second vector making use of a marker gene cloned into a vector of viral origin involves the prokaryotic gene which encodes the enzyme known as **amino glycoside 3' phosphotransferase.** The product of this gene—often called the **neo^R** gene—inactivates the antibiotic **neomycin** which is toxic to ribosomes and therefore kills unprotected eukaryotic and prokaryotic cells. The neomycin analog known as **G418** has the unusual property of being toxic to eukaryotic cells while not affecting the growth of prokaryotic cells. Importantly, G418 can be rendered inactive by the enzyme product of the neo^R gene. Therefore, a vector containing the neo^R gene under the control of a prokaryotic promoter, such as the promoter of the SV40 early genes, can transform a mammalian host cell and rescue it from growth medium which has been supplemented with the G418 antibiotic (Figure 10.6).

CONTROL OF CLONED GENE EXPRESSION IN MAMMALIAN HOSTS

Expressing Cloned Genes in Mammalian Hosts: Enhancers

We saw that a variety of specific sequences of bases appear to play a role in controlling the level of expression of certain genes in both prokaryotic and eukaryotic cells (Chapter Eight). As described, some of the controlling sequences associated with eukaryotic genes are the TATA box and the CAAT box, located approximately 25 base pairs and 75 base pairs, respectively, upstream of the transcriptional start site. In addition, there are enhancer sequences, somewhat less restricted in their spatial relation to their target gene, which function effectively regardless of their orientation within the genome.

Sequences which have similar controlling functions exist in viral genomes as well. For example, a 72-base-pair repeating sequence—also called an enhancer sequence—has been identified in the SV40 genome. This viral enhancer sequence appears to be required for the expression of the SV40 early genes. Experiments show that the viral enhancer sequence, when excised and linked to the rabbit gene which encodes beta-globin and placed on a plasmid cloning vector, induces high levels of expression of the rabbit gene when using the recombinant vector to transform human host cells. In contrast, host cells transformed with plasmids containing the rabbit beta-globin gene without the SV40 72-base-pair enhancer sequence appear to produce no rabbit beta-globin mRNA at all. Similar viral enhancer sequences have been identified in a number of virus particles including Maloney Sarcoma virus, Rous Sarcoma virus, murine leukemia virus, polyoma, and bovine papilloma virus.

Expressing Cloned Genes in Mammalian Hosts: Promoters

Bacterial promoters are sequences of bases which provide a binding site for RNA polymerase, the enzyme required for the synthesis of RNA molecules during the process of transcription (Chapter Eight). Bacterial promoters function with varying levels of efficiency due in part to the presence of specific sequences of bases including the Pribnow Box and the -35 sequence. Variation of the base sequences which make up these constructs generally leads to decreased levels of efficiency in transcription of the associated gene or genes. In addition, some bacterial promoters are inherently **strong promoters,** leading to relatively higher levels of expression of associated genes, while others are **weak promoters.** Examples of strong bacterial promoters include the *lac* operon promoter and a promoter made by fusing a length of the *trp* operon promoter with the *lac* Shine-Dalgarno sequence.

Eukaryotic host cell cloning systems, including mammalian cells, resemble the prokaryotic systems described above in that the expression of their genes often relates to the function of a specific promoter sequence. Thus, manipulation of the promoter sequence in the cloning vector often leads to variations in the level of expression of the gene being cloned. For example, the presence of glucocorticoid hormones induces the expression of **mouse mammary tumor virus** (**MMTV**) genes. When mouse cells are transformed with a cloning vector containing the MMTV genes along with the associated glucocorticoid sensitive promoter, something interesting happens: if the host mouse cell expresses receptors for glucocorticoid hormones and thus responds to these hormones, the MMTV genes are expressed. If, on the other hand, the chosen host cell lacks glucocorticoid receptors thus making it unresponsive to the hormone, the MMTV genes remain unexpressed. A promoter of this type is called an **inducible promoter,** since its function is induced by the presence of a specific molecule or type of molecule.

Another example of an inducible promoter involves expression of the mouse metallothionine gene. Metallothionines, proteins which bind to heavy metals, function in the removal of these poisons from an organism. The presence of heavy metals induces the expression of these genes. A cloning vector containing the mouse metallothionine gene linked to the *E. coli* gpt gene can be used to transform human HGPRT$^-$ cells. An examination of the resulting transformed human cells shows that the expression of the bacterial gpt gene is now induced by the presence of heavy metals. In other words, the metallothionine promoter retains its natural inducible mechanisms and exerts its controlling effects on an exogenous gene—the gpt gene—in a novel environment, the human cell.

Heat shock proteins provide further insight into the control of expression of cloned genes in mammalian host cells. Heat shock proteins, a set of eight proteins, are expressed in ***Drosophila***—fruit fly—cells, as well as in many other eukaryotic cells, when those cells grow at temperatures somewhat higher than normal. The expression of heat shock proteins is under the control of a series of promoters whose functions are induced by heat. Expression of these genes, when introduced into mouse cells which do not normally produce heat shock proteins, continues to be induced by heat. In addition, if the fifteen-base-pair consensus sequence from the heat shock promoters is linked to a gene whose expression is not normally induced by heat, and the resulting vector transferred into a host cell, expression of the non-heat-shock insert gene will now be induced by heat.

DETECTION OF CLONED GENE PRODUCTS ON THE CELL SURFACE

Sometimes normal gene expression results in the production of a polypeptide molecule that is transported through the cytoplasm of the host cell and inserted into the cell's membrane. Such a molecule frequently takes up a position where part of the molecule extends to the exterior of the cell while another part of the molecule projects through the membrane and into the cytoplasm. An example of such a molecule is the **human leukocyte antigen** or **HLA.** This molecule extends through the cell surface of almost all human cells in association with a small polypeptide called **beta$_2$-microglobulin.** Scientists have used the technique of whole-cell hybridization to insert the chromosome containing the human HLA gene into a mouse cell. They then asked whether the HLA gene would be expressed and, if so, whether the gene product would be correctly inserted into the membrane of the host cell. The answer to this question depended on a series of antibodies which had been developed to bind specifically to the HLA molecule, and a complex piece of machinery known as a **fluorescence-activated cell sorter** or **FACS.**

In this experiment scientists Dr. Michael Kamarck and Dr. Frank Ruddle, along with colleagues at Yale University, hybridized human and mouse cells in the presence of the fusogen **polyethelyne glycol** or **PEG**. The resulting hybrid cells initially contained the full complement of human chromosomes along with the mouse genome. As the hybrids matured, however, the human chromosomes randomly segregated from the hybrid cells, resulting in a series of mouse cells which contained varying subsets of the human genome. Scientists then examined the various populations of hybrid cells to determine whether any of them expressed the HLA protein. They reasoned that expression of the protein depended on the retention, within the hybrid cell, of the chromosome containing the human HLA gene. If such expression could be detected, scientists wanted to identify and isolate those cells which expressed the HLA gene (HLA$^+$) from those hybrid cells which lacked such expression (HLA$^-$).

Scientists achieved these goals by treating all of the hybrid cells with an antibody which bound specifically to the portion of the HLA molecule expressed on the surface of the hybrid cell—an **anti-HLA antibody.** The specificity of the antibody allowed it to bind to the HLA molecule present on the surface of any HLA$^+$ cells while preventing it from binding to any other molecule. Thus, HLA$^-$ cells remained unaffected. Another antibody, this one specific for the anti-HLA antibody, was bound to a fluorescent molecule and then allowed to bind to any anti-HLA antibody present on the surface of any of the hybrid cells. Once again, this time due to the absence of anti-HLA antibody, the HLA$^-$ cells remained unaffected. Scientists now faced the task of separating the fluorescent cells—those expressing the HLA antibody bound to the HLA protein—from the non-fluorescent cells. To accomplish this difficult task scientists turned to the FACS, a machine capable of separating single cells from heterogeneous populations of cells. Analysis of the total population of hybrid cells showed that approximately fifteen percent of them expressed the HLA gene as determined by the presence of anti-HLA antibody. Once that fifteen percent had been separated from the remaining eighty-five percent of non-expressing cells, researchers cloned the smaller population and examined it for human DNA content. Analysis eventually showed that the presence of the HLA molecule on the hybrid cell surface correlated with the presence of human chromosome 6 in the hybrid cell. Thus, these experiments showed that a host cell could be transformed with a gene encoding a protein which could, in turn, be properly transported through the cell and expressed on the cell surface. In addition, analysis of such transformed hybrid cells allowed the human HLA gene to be assigned a chromosomal map position on human chromosome 6. (Chapter Eleven).

IN DETAIL: THE FLUORESCENCE-ACTIVATED CELL SORTER

A complex machine, the FACS (Fluorescence-Activated Cell Sorter) uses the immunological interaction of antigens and antibodies to sort populations of heterogenous cells. Originally developed along the lines of a prototype high-speed ink jet printer—the machine broke a stream of ink into droplets which were electrically charged and then deflected to appropriate positions on the paper—the cell sorter has not only developed into the modern ink jet printers of today, but has also become invaluable in scientific research. Almost unique in its capabilities, the FACS takes a mixed population of cells, detects within it a homogeneous subpopulation of cells, and then actually separates the few homogeneous cells of interest away from the larger parent population.

The FACS detects and separates cells on the basis of how much fluorescent dye is bound to individual cells. A heterogeneous population of cells is first treated with the appropriate antibodies and fluoresceinated compounds as described in the text above. This treatment results in the selective labelling or tagging of target cells. Non-target cells remain untagged. The treated cells—both tagged and untagged—are then mixed with a small amount of growth medium or other liquid to produce a liquid **cell suspension.** The cell suspension, then forced under pressure through a series of hollow tubes until it reaches a nozzle, continues on its way through a hole in the tip of the nozzle, generally between 50 and 90 microns in diameter. The cell suspension emerges from the nozzle tip, positioned into the center of a stream of **sheath fluid.** Sheath fluid, usually physiological saline—0.9% sodium chloride—serves to ensure that the central sample stream of cell suspension maintains a relatively uniform size and shape, and it holds the sample stream in a relatively constant central position. As all of this takes place, the entire nozzle assembly vibrates at a frequency of approximately twenty thousand cycles per second. This vertical vibration breaks the sample stream and surrounding sheath fluid into about twenty thousand uniform droplets per second. A strobe light, whose frequency matches the nozzle vibrational frequency, allows visualization of the individual droplets which begin to form at a measurable distance downstream of the nozzle tip.

Focused at a point within the pre-droplet stream is a beam of laser light whose operating wavelength causes excitation of the fluorescent label on tagged cells within its path. Fluorescent dyes, when exposed to light of particular wavelengths, give off energy. This energy, called a **fluorescent emission,** focuses onto a **photomultiplier tube** which in turn generates an electric signal proportional in size to the amount of energy emitted by the analyzed cell. In addition, as any cell passes through the laser beam an amount of light, roughly proportional to the cross-sectional area of the cell, is scattered out of its undisturbed path. This **scattered light** registers as a second electric signal upon which the FACS is triggered to respond. The two signals produced—the fluorescent emission which indicates the presence of fluorescently labelled molecules on the cell and the scattered light which is a rough indicator of cell size—are compared to determine the possible presence of a cell with the desired level of fluorescence which indicates the presence of the target molecule. When such a cell is detected, the FACS produces an electric pulse. This pulse puts an electric charge onto the sample stream just as the tagged cell is being included into a forming droplet. On termination of the electric pulse the pre-droplet stream loses its charge while the newly formed droplets retain theirs. In this manner, droplets containing fluorescently tagged cells are electrically charged, while both fluorescently empty droplets and the pre-droplet stream remain electrically neutral. Droplets containing cells expressing a level of fluorescent emission below a desired threshold can also trigger an electric pulse. Such cell-containing fluorescently negative droplets receive an electric charge opposite to that received by fluorescently positive droplets.

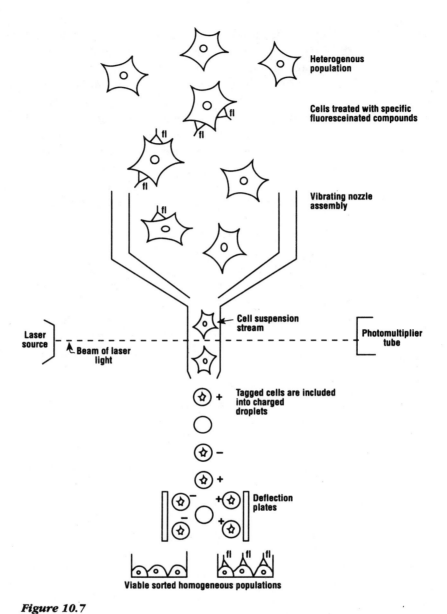

Figure 10.7
The Fluorescence-Activated Cell Sorter (FACS) allows identification and isolation of homogeneous subpopulations of cells from heterogeneous populations. This schematic diagram illustrates basic functioning of the FACS.

All droplets, regardless of charge, are directed from the nozzle downward between two electrically charged plates. The constant electrostatic field between these plates causes charged droplets to deviate from the path of the main stream. Droplets of like charge, encountering a uniform electrostatic field, migrate in a path predictably deviant from the undeflected path. Droplets with opposite electric charges migrate in opposite directions. Both positively and negatively charged droplets can therefore be collected in separate containers as they leave the mainstream. Cells contained in deflected droplets often retain viability and can be expanded in tissue culture for further examination.

REVIEW AND SUMMARY

1. Despite the presence of the appropriate controlling sequences in a variety of cloning vectors, a number of higher eukaryotic genes are not expressed when cloned into either bacterial or yeast host cells. For this reason, mammalian cells are sometimes chosen as host cells. Although mammalian host cells pose no danger of bacterial contamination of the cloned gene product and permit thorough analysis of the patterns of mammalian gene control, they have certain drawbacks: these eukaryotic host cells permit the growth of various mammalian pathogens which pose a potential danger to any human patients treated with therapeutic molecules produced using this type of cloning system.

2. A variety of techniques can lead to the introduction of exogenous DNA into a mammalian cell. One of the simplest, **somatic cell hybridization,** results when cells of two or more different species are caused to fuse with one another by the application of a fusogenic agent such as PEG or Sendai virus. Hybrid cells resulting from this process tend to randomly segregate the chromosomes of one of the parent cells. The genomes of such cells are, therefore, unstable and not suitable to many types of transformation experiments.

3. Mammalian cells can be transformed with exogenous DNA by first mixing the donor DNA with a source of excess calcium ions and then applying the resultant **DNA/calcium precipitate** to host cells. Some of the cells incorporate this precipitate through **endocytosis.** The incorporated DNA either immediately degrades, degrades after a period of time, or integrates into the host genome. The fate of the integrated DNA directs the type of expression—absent, transient or stable—of the exogenous DNA. This inefficient technique results in a relatively small percentage of successfully transformed cells.

4. Donor DNA can be **microinjected** into the pronuclei of fertilized mouse eggs to create transgenic mice. The resultant embryos can then be implanted into a host mouse uterus and brought to term. When born, some of these mice continue to express the foreign gene in a mosaic pattern. As such, not all of the tissues of a single animal contain the foreign gene, nor will all of the cells of a positive organ be positive. Experiments show that products of microinjected genes can be expressed in the full-grown mouse and can also have obvious phenotypic effects. Similar microinjection studies have also involved other animals including pigs and sheep.

5. Transformed host cells are often detected against a background of untransformed cells with the use of **selective media.** Such growth media contain various chemicals which interact with certain gene products to enhance or inhibit cell growth. An example of selective media, **HAT medium,** has chemicals which prevent the growth of cells unable to synthesize the TK enzyme. Therefore, successful transformants resulting from the treatment of TK⁻ host cells with a cloning vector which contains a TK gene can be detected by their ability to grow in HAT medium.

6. TK acts as a marker for non-selectable DNA sequences when the two types of molecules are coprecipitated prior to transformation of host cells. Although the two gene types are not physically linked before they are introduced into the host cell, the presence of the TK gene in the host cell frequently indicates the presence of the non-selectable gene as well. Human DNA has a natural marker sequence known as **Alu.** This repetitive sequence of DNA occurs so many times in the human genome that its repeated presence in a sample of DNA identifies that DNA as being of human origin.

7. **Microinjection** involves the injection of DNA into a host cell via a mechanically controlled microneedle. This valuable technique results highly efficient transformation and does not require the presence of a selectable marker in the exogenous DNA.

8. Some vectors of viral origin have been engineered to include a selectable marker gene. A variety of such vectors has been isolated from **SV40,** a small tumor virus which normally infects monkey cells. The two major engineered SV40 vector types differ in that one contains a bacterial **gpt gene** while the other contains a bacterial **neo^R** gene as a selectable marker. Each of these vectors is detectable in appropriate selective media and can carry an insert of donor DNA to be cloned. A **helper virus,** often used with engineered SV40–based vectors, provides any missing gene products when the cloning vector cannot direct synthesis of these missing products due to the presence of the donor DNA insert.

9. The expression of cloned genes in mammalian host cells depends on the presence of the appropriate controlling sequences. Such sequences, including **enhancers** and **promoters** of a variety of origins, are often included in the cloning vector itself. Experiments show that the presence of an inducible promoter on a cloning vector often renders expression of the normally non-induced donor gene inducible as well.

10. The presence of a cloned gene product on the surface of a host cell can be detected with a **Fluoresence-Activated Cell Sorter.** This machine uses a combination of immunological reagents and laser light to detect the presence of single cells which express particular molecules on their cell surfaces. Once detected, the FACS also has the ability to separate that single cell from the mass of heterogeneous cells which do not express the target molecule. That cell can be grown in tissue culture until it divides and re-divides to form a homogeneous population—expressing the target surface molecule—which can then be investigated and examined.

Applications of Biotechnology: Human therapeutics

Most Americans at some point perceive themselves to be heavier than they would like, so they restrict their caloric intake by going on a diet. Traditionally, dieters eat diet food—cottage cheese, celery, carrots, limited amounts of plain fruit, and great quantities of diet soda. Within the past few years, however, dieters have seen a radical expansion in the variety of low calorie foods available to them. With the advent of artificial fats and new artificial sweeteners—dieters now enjoy a wide range of foods including low-calorie ice cream, candy, bakery products, and treats and beverages of all sorts. Although these products can be consumed safely by most people, they have potential side effects; in fact the labels on many of these low-calorie alternatives display a warning sign that says "Phenylketonurics: contains phenylalanine" (Figure 11.1). This warning applies to products flavored with an artificial sweetener, **aspartame,** which contains large quantities of the amino acid **phenylalanine.** Aspartame, extremely sweet in taste, is composed primarily of an ingredient that occurs naturally at low levels in many foods and, most importantly, shows no measurable negative side effects on the vast majority of individuals. There is, however, one important exception.

One in 11,000 people born in the United States have a disease called **phenylketonuria** or **PKU.** People affected by this inherited disease are not able to synthesize an enzyme called **phenylalanine hydroxylase.** Normally this enzyme functions to convert phenylalanine into another amino acid known as **tyrosine.** In the disease phenylketonuria, however, the absence of this enzyme leads to an abnormal build-up of phenylalanine in the body. This store of accumulated phenylalanine then converts into a class of chemical compounds known as **phenylketones** which, in turn, causes abnormal development of nervous tissue in the brain. PKU, if allowed to progress unchecked, leads to progressively increasing levels of seizure activity, severe mental retardation, and a shortened life span. However, the progression of PKU can be halted by providing patients suffering from this disease with a diet which completely excludes phenylalanine. Therefore, people with PKU should not consume products containing aspartame.

Applications of Biotechnology: Human Therapeutics

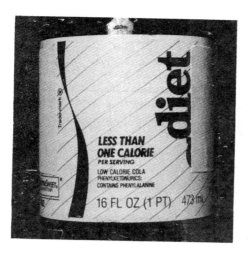

Figure 11.1
Many modern grocery products are sweetened with a synthetic sweetener known as Nutrasweet, composed primarily of the amino acid aspartame.

The incidence of mental retardation caused by PKU has dropped dramatically in recent years due to sufficiently early dietary intervention, which has only been possible since the development of genetic engineering techniques which have allowed the detection in newborn infants of the mutant gene which causes this disease. These detection techniques revolve around the use of genetically engineered gene probes and a knowledge of the genetic bases of various human diseases. This chapter discusses some of the aspects of the complex relationship between biotechnology and human health.

OVERVIEW

Perhaps no area of modern existence has been so greatly affected by biotechnology as that of human therapeutics and human health care. Diseases such as PKU meant severe retardation and eventual early death just twenty or thirty years ago; now they are treated in a relatively routine fashion. Pregnancies that once resulted in the birth of babies with inherent genetic defects can now be detected and sometimes even treated during the months of the first trimester. Diseases such as hepatitis, for which the only prevention was once avoidance, can now be prevented with a genetically engineered vaccine. These are just a few examples of the many ways in which biotechnology impacts our lives in the modern world.

This chapter examines more closely some areas of modern life that have been so greatly affected by the science of biotechnology. It also answers the following questions:

1. What is the Human Genome Mapping Project and what effect will it have on health care?
2. How close are we to the therapeutic uses of gene replacement therapy?
3. What is a magic bullet and how can it help cancer patients? What about cystic fibrosis patients?
4. Is there a drug which acts as "cardiac drain cleaner?"
5. What is a subunit vaccine and does it have any advantages over a whole-virus vaccine?

HUMAN GENETIC ABNORMALITIES

Genetic Abnormalities Are Common

Estimates suggest that approximately half of all human pregnancies end in miscarriage or in spontaneous abortion, many of which occur before the woman even knows herself to be pregnant. Of those fetuses which spontaneously abort at some time during the first trimester of pregnancy, about half show chromosomal abnormalities. Even when a pregnancy results in a live birth, the game of genetic roulette is not yet over. An estimated ten percent of all newborn babies have a genetic abnormality significant enough to result in the development of a genetically based disease sometime during the life span of that individual. About half of the ten percent of affected individuals are actually born with the genetic defect already in evidence.

Two Levels of Mutation in Genetic Disease

Researchers have long realized that the heritable nature of genetic diseases must stem from the involvement of a mutant gene or genes as the causal factor: that mutant gene must produce a polypeptide product which is abnormal in either structure, function, or quantity. Scientists, therefore, have a two-fold task. First, they must identify the abnormal gene product which plays a role in any particular disease state. Second, they must identify the precise chromosomal location of the mutant gene which encodes the abnormal gene product. Both of those tasks have been greatly aided by the increasingly sophisticated techniques of biotechnology.

The Genetic Cause of Many Inherited Diseases Is Unknown

Almost 4,000 genetic diseases had been described by scientists and physicians prior to the year 1990. Although these 4,000 diseases are clearly inheritable and therefore must result from a genetic defect of one sort or another, both the defect itself and the chromosomal map position of the responsible gene remain unidentified for the vast majority of inherited human diseases. This results in part from the fact that experiments used in non-human animal species to gain such information are impossible to carry out in humans because of the long generation time of humans and—even more importantly—due to the ethical issues involved in directing the mating patterns of individual human beings. So, most genetic diseases have been associated with a specific defective gene product or chromosomal map position through an examination of the pattern of disease inheritance across multiple generations of the same family. This kind of information is called **pedigree analysis.** Knowing the rules of Mendelian inheritance along with detailed familial pedigree information, researchers can sometimes connect a gene, or a chromosome at the very least, to a particular disease, especially when the faulty gene lies on a sex chromosome rather than on an autosome. Unfortunately, this type of analysis is hampered by the fact that many genetically based diseases, including schizophrenia, alcoholism, and heart disease, appear to be caused by a group of two or more genes acting in concert with one another. Such disease-causing genetic interactions remain especially difficult to identify by the traditional techniques of genetic examination in humans—specifically pedigree analysis.

Despite the difficulties inherent in pedigree analysis, this technique has proven useful in the examinations of some inherited diseases. Color-blindness and hemophilia, the first genetic abnormalities to be **mapped**—assigned to a gene at a specific chromosomal location—were, in fact, assigned using the technique of pedigree analysis. In the early part of the 20th century, medical doctors and other scientists realized that the patterns of inheritance of these diseases simulated the pattern of inheritance for sex and the X chromosome. This insight led scientists to make the assertion—later shown to be correct—that the genes responsible for causing these two disease phenotypes must be located on the X chromosome. However, it was not until the 1980s that the gene products associated with these diseases—Factor VIII for hemophilia and the visual pigment genes for color blindness—were identified and the responsible genes precisely located on the X chromosome and later sequenced and cloned.

ABNORMAL PROTEINS ARE IDENTIFIED:

Metabolic Disorders

Metabolism is the name given to groups of chemical reactions that convert **substrates**—raw materials—into substances necessary for the proper functioning of an organism. A particular enzyme drives or enhances each of the chemical reactions that makes up a metabolic pathway. If that enzyme is either absent or defective in some way, the chemical reaction either slows down to an ineffective rate or even halts entirely. As a result, a particular substrate is not converted into the necessary materials and functioning of the organism is compromised.

Approximately three hundred human diseases which are due to the presence of known inherited metabolic disorders have been identified. For each of these diseases, the lack of an enzyme or enzymes causes the particular disease phenotype, such as phenylketonuria as described above. Other so-called **metabolic diseases** include **Tay-Sachs disease** in which the enzyme **hexosaminidase A** is absent, and **xeroderma pigmentosum** in which the DNA repair enzymes are missing. The absence of hexosaminidase A in Tay-Sachs disease results in developmental and motor deterioration beginning sometime in the second half of the first year of life. This deterioration eventually leads to the loss of sight and hearing, as well as to neurological seizure activity which becomes progressively worse with time. Children affected with Tay-Sachs Disease have a life expectancy of approximately five years. The absence of the DNA repair enzymes in xeroderma pigmentosum leads to a marked sensitivity to the ultraviolet

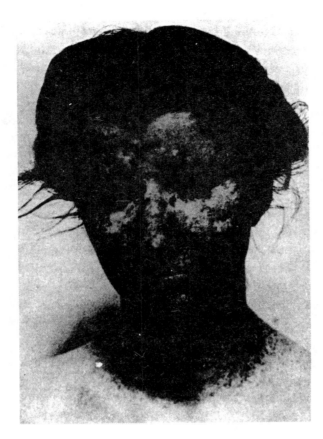

Figure 11.2
Xeroderma pigmentosum is a metabolic disorder in which DNA repair enzymes are nonfunctional. As a result, patients develop multiple skin cancers.

rays of sunlight. This sensitivity occurs because of the body's inability to repair the DNA damage inevitably caused by exposure to ultraviolet light. As a result, multiple skin cancers, including malignant melanoma, usually occur prior to twenty years of age in victims of this disease (Figure 11.2).

ABNORMAL GENES ARE IDENTIFIED:

Mapping with Somatic Cell Hybrids

Since the mapping of color blindness and hemophilia in the early 1900s, genes were mapped at the rate of a few per year for many years. That changed in 1967 when scientists first used the technique of somatic cell hybridization (Chapter Ten) to assign the gene which encodes the enzyme **thymidine kinase** to a map position on human chromosome 17. The development of the **clone panel**—a series of interspecies somatic cell hybrids, each of which contains a different subset of the human genome—made the technique of mapping by somatic cell hybridization somewhat more versatile (Figure 11.3). By the late 1980s approximately 50 genes had been mapped to their appropriate human chromosomal locations using these techniques. Recently additional techniques have been developed which allow examination of individual chromosomes in greater depth.

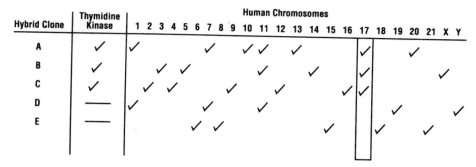

Figure 11.3
An example of genetic mapping using a clone panel. Hybrid clones are examined both for their expression of the target gene—in this case thymidine kinase—and for the subset of human chromosomes which they contain. The target gene can only be located on a human chromosome whose appearance matches the expression pattern of the gene. Here, both thymidine kinase and chromosome 17 are present in clones A, B, and C, and are absent from clones D and E. The appearance of no other chromosome matches expression of the target gene. The gene which encodes thymidine kinase must therefore lie somewhere along the length of chromosome 17.

Mapping with Sorting Chromosomes

Using the FACS (Chapter Ten) scientists can actually separate human chromosomes into separate piles based on their relative physical characteristics. As described in Chapter Ten, FACS analysis relies on the interaction of immunologically specific labelled antibodies with energy from a beam of laser light. By treating chromosomes with a fluoresceinated **anti-DNA antibody**—an antibody which binds specifically to DNA—researchers uniquely label each chromosome based on the amount of DNA it contains. Larger chromosomes, containing relatively greater amounts of DNA, are fluorescently brighter while smaller chromosomes will be fluorescently dimmer. Since each pair of human chromosomes contains a unique amount of DNA, this procedure ideally results in individual piles, each consisting solely of homologous chromosomes. In other words, one pile should contain only copies of human chromosome 1, another contains only copies of human chromosome 14, and so on. In practice, chromosomal separations are somewhat less efficient and often result in the mixing of two or three types of chromosomes. Even this level of separation is, however, useful in chromosome analyses as described below.

This separation of chromosomes, achieved by using a FACS to sort labelled chromosomes, resembles sorting a population of cells based on the presence or absence of particular cell-surface molecules. Labelled chromosomes, forced through the vibrating nozzle of the FACS, intersect a beam of laser light. Fluorescently labelled chromosomes give off an amount of energy which is proportional in amount to the quantity of label, and therefore to the quantity of DNA, in the chromosome. Some of the forming droplets include chromosomes and receive an electrical charge based on their relative amount of fluorescence. All of the droplets pass through an electrostatic field and all droplets of like charge are deflected through the field in the same manner. Thus, all droplets of like charge collect in individual vessels or as individual spots on specially treated collection paper; consequently, chromosomes can be sorted into individual containers based on their identity or onto separate spots on a nitrocellulose filter. Under optimal conditions and using a flow cytometer equipped with two lasers, each tuned to a different wavelength, it is possible to sort between five and six million chromosomes in a single hour. Interestingly, it takes approximately fifty thousand chromosomes of a single type to provide material sufficient for base sequence analyses. This homogeneous preparation of chromosomes can be analyzed with labelled complementary nucleic acid probes to ascertain the presence of certain specific genes or DNA sequences using the technique of Southern blot analysis (Chapter 5).

Mapping with Known Genetic Markers

A great deal of genetic mapping takes place by the identification of any possible associations between genes of unknown map position and genes whose map position is known. In other words, if the latter type of gene—known as a **marker** because it identifies or marks a known chromosomal location—most often appears with another gene, the second gene probably has a map position somewhat close to the marker gene. For example, if people with blue eyes always have freckles, and if freckles never appear on a person who didn't also have blue eyes, we might conclude that the gene for blue eyes and the gene for freckles are located very close to one another on a single chromosome. If, however, the presence of blue eyes was completely independent of the presence of freckles, we would conclude that the genes were not physically close to one another. Therefore, if our previous research ascertains that gene M is located on chromosome C, and we know that gene N always appears with gene M, we can conclude that the genes are **linked** and that both genes are located on chromosome C.

Markers and Probes Can Be Used for Genetic Fishing

Currently, markers have been identified for a wide variety of chromosomal regions. All human chromosomes contain at least one identifiable marker; in fact, more than a thousand genetic markers are distributed at intervals of approximately 10 million base pairs throughout the human genome. This remarkable array of markers has made many areas of the human genome accessible to research by providing a known spot that can be identified by the binding of a complementary probe. If the human genome is reduced to fragments by the action of restriction endonucleases, the use of markers and their complementary probes allows a researcher to fish out a particular fragment by isolating only that fragment which binds to the probe. By examining the non-marker DNA also located on that fragment, scientists can identify other genes in the neighborhood of the original marker. This resembles the mapping-by-phenotype, described in the previous paragraph, but with important differences: the fishing technique allows the direct examination of genes rather than an indirect examination of those genes via their resulting phenotypes. For the first time, non-expressed genes can be examined. This advance led to some interesting dilemmas: suddenly scientists encounter the dilemma of a gene whose function is unknown; previously too many functions had unknown genes!

Chromosome Walking: Mapping Piece by Piece

Sometimes an area of interest resides somewhere between two markers. Although two known genetic markers might flank a particular area, that area may remain inaccessible due to the limitations of cloning protocols: even if we fished out the entire fragment using a complementary probe, we could not clone it and generate sufficient experimental material. Unfortunately, the several hundred thousand to as many as five million base pairs or more that often separate a marker from an area of interest cannot be incorporated into a single cloning vector. This means that the target gene cannot be isolated by making a library and screening that library with a probe which is complementary to the marker. Instead we must rely on a technique known as **chromosome walking.** In this technique scientists assemble a library containing fragments of the entire genome. A probe, constructed from a small length of DNA isolated from one end of a known marker sequence, is then used to screen the library in an attempt to isolate a transformant which contains the end of the marker-containing fragment along with the next length of the chromosome. A short DNA sequence located close to the end of the new overlapping sequence is then isolated from that clone and used to re-screen the library for a transformant which contains the new probe along with the next length of chromosome. This technique is carried out repeatedly until the chromosome has been walked in small overlapping steps from the original marker to the area of interest.

RFLPs and Gene Mapping

One way in which biotechnology impacts human health has been through the use of restriction endonucleases to cleave lengths of DNA. As described in Chapter Four, the term **restriction fragment length polymorphism,** or **RFLP,** describes the DNA fragments of varying lengths which occur when homologous DNA, isolated from two or more individuals, is treated with the same restriction endonuclease. This size variation results from the variation in bases which can be found in the non-coding regions of the DNA of different individuals. Since these variations in base sequence—known as **site polymorphisms**—may affect one or more restriction endonuclease recognition sites, DNA from different individuals treated with the same restriction endonuclease most likely yields DNA fragments of different sizes—RFLPs.

The RFLPs resulting from site polymorphisms have two characteristics which make them especially useful in gene mapping: 1) they are constant within an individual and pass from generation to generation as each individual provides a genetic complement to a gamete which goes on to form a zygote. 2) Multiple RFLP-causing site polymorphisms, scattered liberally throughout the human genome, make it likely that at least one heritable RFLP appears in the vicinity of any particular gene. For these reasons RFLPs can be considered as another form of phenotypic variation, just like blue or brown eyes, and can be used as markers which aid in the examination of the human genome. By examining the DNA of many members of the same family and analyzing the relationship of various RFLPs with the presence of a disease state, scientists can sometimes conclude that the presence of a particular disease often occurs with the presence of a particular RFLP. This type of information suggests that at least one gene involved in the origins of the disease in question is located in the vicinity of the RFLP on the chromosome.

RFLPs proved especially useful in mapping certain disease genes such as that which leads to the development of **Huntington's disease**—a late onset degenerative disease. Generally not manifesting itself until the patient is between thirty-five and forty years old, Huntington's disease causes the progressive degeneration of the central nervous system. Patients with this disease generally die within a few years of the initial onset of their symptoms.

Scientists led by Drs. Nancy Wexler and James Gusella of the Hereditary Disease Foundation and the Massachusetts General Hospital, respectively, analyzed the DNA of more than five thousand people living in Venezuela, all of whom descended from a single woman who died of Huntington's disease almost one hundred years ago. Using this information, Drs. Wexler and Gusella and their colleagues correlated the presence of a particular RFLP, caused by a site polymorphism located at one end of human chromosome 4, with the eventual development of Huntington's disease. Examples of other genes which have been mapped using this type of RFLP analysis include the gene for cystic fibrosis which has been assigned to chromosome 7; the gene for familial Alzheimer's disease, chromosome 21; the gene for retinoblastoma, chromosome 13; and the gene for sickle-cell anemia, chromosome 11.

All of the types of mapping mentioned above revolve around the fact that something is known about the gene to be mapped. That means that we can work backwards. Perhaps we know the identity of the defective enzyme in a particular disease, or perhaps we isolated a particular chromosome and are looking for products of genes residing on that chromosome. Maybe we've analyzed the morphology of the chromosomes of a patient with a particular disease and have discovered a structural abnormality. In any case, we have some information and proceed to work backwards.

What if it were possible, instead, to generate a sort of road map of the human genome? What if we could identify every single gene on every human chromosome? This notion, once belonging to the realm of science fiction, is rapidly appearing more feasible due to great progress in the techniques of biotechnology.

Mapping the Human Genome

Many scientists feel that a detailed map of the entire human genome would prove to be an invaluable and beneficial tool for human health care and diagnostics. This goal has been formalized into an international effort called the **Human Genome Mapping Project.** Hundreds of scientists, spending approximately three billion dollars over the fifteen years between 1990 and 2005, are attempting to determine the precise sequence of all of the approximately 3 billion base pairs which make up the 23 pairs of human chromosomes.

The project, directed by Nobel laureate Dr. James D. Watson who heads the National Center for Human Genome Research at the National Institutes of Health, intends to untangle the amazing intricacies of the human genome. As we have seen, the human genome consists of about 3 billion base pairs arranged into 23 pairs of chromosomes, each consisting of about 150 million base pairs. Those base pairs, in turn, contain somewhere in the neighborhood of 100,000 genes. Only about five percent of the human genome appears to be in the form of genes with specific protein products. The remainder of the DNA appears to be filler DNA whose precise function remains unknown. According to Victor McKusick's *Mendelian Inheritance in Man*—a publication which is updated every five years and which describes advances in the mapping of human genes—only about 4,500 of those 100,000 genes have been identified to date. Even worse, only about 1,500 of those genes have been assigned approximate locations on a human chromosome.

The three primary goals of the human genome mapping project include:

1. identifying many more genetic markers so that no area of the human genome is far from such a known marker.
2. assigning all of the genes of the human genome to a chromosomal map position.
3. ultimately sequencing the entire set of human chromosomes.

According to the 1988 National Research Council report, the data from the sequencing project alone will fill a million page book with a double string of letters representing the base pairs of the human genome!

The potential benefits of the information gained through the human genome mapping project are enormous. In the area of medical research scientists hope that the information gained will enable them to develop diagnostic tests for a wide variety of human diseases, to identify therapies and treatments based on an understanding of the genetic defects active in many human abnormalities, and to play a role in the prevention of disease by an understanding of genetic factors which predispose certain individuals toward the development of particular diseases. In addition, the human genome mapping project will undoubtedly play a role in answering many questions of basic research. For example, this information may enable us to understand the presently indecipherable role of the so-called junk DNA of the genome, and may provide insight into the evolutionary process which led to modern day man.

Although it is indisputable that great volumes of data will be gathered by the scientists involved with the Human Genome Mapping Project, other scientists question the value of this colossal project. These critics argue that the project will generate data which, with present day techniques, is essentially useless—a suspected ninety-five percent of the human genome has no known useful function. Some scientists also point out that the time and expense required for a detailed analysis of the human genome are enormous and may take assets away from other important research projects. In addition, a knowledge of the base sequence of the human genome may not bring any clearer understanding of the genetic function or relation to disease, of most sequences.

Examining other aspects of this project, ethicists express concern about the impact which detailed genetic knowledge could have on our lives. It may someday be possible to detect hereditary factors ranging from a genetic predisposition toward alcoholism to the presence of a gene correlating to the development of a late-onset cancer. Should potential employers have access to this type of information? What about insurance companies? Does the potential availability of this type of information signal the onset of what has been called an "era of genetic bias"?

HUMAN HEALTH CARE

Gene Replacement Therapy

The highly effective combination of biotechnology and the human genome mapping project will most likely result in knowledge of the precise base sequence of many, if not all, of the active human genes. Many scientists believe that this information, coupled with the ability of researchers to either isolate intact genes from the genome or to synthesize genes in the laboratory, will make it possible to cure genetic disease by replacing the mutant gene with a non-mutant gene copy. Such a scenario is called **gene replacement therapy.**

In gene replacement therapy a faulty gene is identified from the one hundred thousand or so genes that make up the human genome. A normal copy of that gene would then be cloned or sequenced and synthesized. The replacement gene would then be placed in the appropriate **stem cells**—cells which give rise to differentiated cells during development—of the affected individual. As those stem cells develop into mature differentiated cells they carry with them a copy of the corrected gene. Ideally, the replacement gene will begin to function at the appropriate stage of cell development, expressing the proper levels of the missing gene product or products and alleviating the disease condition. To reach this goal, the inserted gene would have to be expressed under the appropriate level of control so that the product would be present in the right amount. Further, the gene would have to be incorporated into a location on the host genome where it would not damage the functioning of any host gene, and the cells containing the normal gene would have to divide rapidly enough to overtake the growth of the cells containing the mutant gene.

The first semi-successful example of this type of gene replacement therapy took place in the early 1980s. At that time Dr. Richard Palmiter and his colleagues attempted to cure "little" mice. This strain of mouse suffers from a congenital lack of mouse growth hormone. As a result, "little" mice are, in fact, little. By microinjecting "little" mouse embryos with the rat growth hormone gene, researchers hoped to cure the genetic defect causing the abnormally small mice. What happened was extremely interesting. The rat growth hormone gene became functional in some of the mice embryos. Those embryos developed into full-grown mice that achieved a size which was larger than the average "little" mouse. Unfortunately, however, the inserted rat growth hormone gene was not under appropriate genetic control of expression—the microinjected animals grew and grew, eventually producing animals as much as one and one-half times the size of a normal mouse! While this experiment marked the first effective attempt at the modulation of a genetic abnormality by gene replacement therapy, it also reinforced the need to maintain genetic function under appropriate control mechanisms.

Gene Therapy: Germ vs Somatic Cells

Two possible cellular targets for the process of gene replacement therapy include the somatic cell and the germ cell. The choice of target cell type has important implications in the result of the experiment. A new gene, placed into a somatic cell of an animal, often results in an altered phenotype for that cell and potentially for the animal as a whole. That inserted gene, however, never passes on to the offspring of the animal. Although a particular phenotype has been altered in the animal, reproductive genotype remains unchanged. In contrast, when a new gene is inserted into a germ cell, the animal may experience alterations in both phenotype and reproductive genotype. The offspring and all future generations descended from that animal may also express the new inserted phenotype. As a result, the gene pool of that species will have been altered through human experimentation.

Such a possibility raises the ethical issues involved with the thought of **eugenics**—the practice of deliberate manipulation of the human genome to produce a particular type of offspring. If we can put ideal genes into a fertilized human egg, and if the resulting human can pass those genes onto its offspring, then we have made an artificially induced change in the human gene pool. For this reason many people feel that gene replacement therapy is morally and ethically wrong. Others believe that gene replacement therapy should be divided into two categories and that manipulations involving somatic cells should be considered to be no different from the more traditional medical therapies since no permanent alteration of the germ line is introduced. Whatever your individual point of view, remember that each new development in scientific research meets with skepticism—Galileo was jailed as a heretic for daring to suggest that the earth was not the center of the universe. As time progresses and knowledge expands, attitudes change and things once considered morally unacceptable become more and more a part of normal life. We don't know how future generations will view the morality of gene replacement therapy, but attempts to halt research or suppress knowledge have never been entirely successful.

Figure 11.4
David was born in Texas in 1971 with Severe Combined Immune Deficiency. SCID left David without a functioning immune system and meant that he had to live in a sterile plastic bubble until his death from complications of a bone marrow transplant in 1984.

Somatic Cells as Targets for Gene Therapy

In theory, gene replacement therapy involves the insertion of a nonmutant form of the affected gene into an appropriate stem cell so that all descendants of that cell contain the correct gene as would have occurred in normal development. In practice this is not generally feasible due to the relatively small numbers of stem cells and the difficulty in their precise identification. In an attempt to circumvent this problem scientists such as Dr. R. Michael Blaese, W. French Anderson, and their colleagues at the National Institutes of Health (NIH) transform blood cells called **lymphocytes** with normal genes. They then insert these transformed cells into enzyme-deficient patients by transfusion with the hope that the engineered cells will provide the missing enzymes.

Drs. Blaese and Anderson have been working with children affected by a rare disease known as **adenosine deaminase deficiency** or **ADD**. ADD, a severe disorder of the immune system, occurs once in every 200,000 live births. ADD gets its name from the enzyme **adenosine deaminase (ADA)** which is necessary for the production and maintenance of two of the major cells types active in the immune response. Without these cell types—B lymphocytes and T lymphocytes—affected children are susceptible to every infection to which they are exposed. ADA deficient children, whose immune dysfunction strongly resembles that of David, the "bubble boy" from Texas (Figure 11.4), die at a very young age from the multiple infections which run wild throughout their bodies as a result of their non-functioning immune systems.

The ADA gene, which has been assigned to a map position on human chromosome 20, consists of 32,000 base pairs. Analysis of this gene shows that it contains 12 exons which, when taken together, account for only about five percent of the total number of bases in the gene. The remaining ninety-five percent of the bases are found in non-coding intron sequences. In preliminary experiments Drs. Blaese and Anderson have already been able to insert the human ADA gene into B and T lymphocytes isolated from ADD children. These engineered cells, viable under laboratory conditions, produced potentially therapeutic levels of the ADA enzyme.

In September 1990 Drs. Blaese and Anderson took their exciting research one step farther by isolating T cells from the blood of a four-year-old girl with ADD. They transformed those cells with a genetically engineered cloning vector containing a copy of the nonmutant ADA gene. Experimental analysis showed that approximately ten percent of the treated cells were producing the ADA enzyme. The engineered T cells, containing the inserted ADA gene, were then returned to the child by the simple technique of blood transfusion. The researchers predict that the newly inserted gene will function properly, synthesizing sufficient ADA to restore normal immune function to the affected child. Only the passage of time will show us the efficacy of this type of treatment, for ADD as well as for other disease states.

HUMAN THERAPEUTIC MOLECULES

One of the earliest uses recognized for the products of biotechnology was the potential of therapeutic molecules. Recombinant DNA techniques have allowed the production of seemingly countless types of molecules which play a role in the maintenance of good health or in the alleviation of ill health. These molecules can be divided into categories based on their general type and usage.

Hormones Made by Recombinant DNA Techniques

Hormones are proteins that are produced by the endocrine glands of the body. Hormones act as chemical messengers, bringing information from endocrine glands, through the blood supply, to remote areas of the body. This information results in the modulation and regulation of various physical processes ranging from growth and reproduction to digestion and the level of sugar in the blood, to name but a few. When specific hormones are absent or are present in abnormally low amounts, physical pathologies can result. These pathologies can often be remedied by the administration of exogenous sources of the missing hormone. Unfortunately, it is not always easy or possible to generate adequate amounts of pure hormone for use in treatment protocols. For this reason, scientists have been working on the cloning of a large number of human hormones including **human growth hormone** and **insulin.**

HUMAN GROWTH HORMONE Congenital dwarfism results when children are born with pituitary glands which are unable to synthesize adequate levels of **human growth hormone** or **hGH.** People lacking this hormone attain maximum growth levels of approximately 4 feet. Prior to the advent of genetic engineering technologies, such individuals were treated with the administration of hGH which had been isolated from the pituitary glands of human cadavers. This type of treatment, however, required hormone from as many as fifty cadavers to treat a single hGH-deficient child for one year. Scientists have largely remedied this lack by inserting the human growth hormone gene into *E. coli* host cells. Those engineered cells then go on to make large quantities of the required hormone under laboratory conditions. The genetically engineered hormone goes by the name of **protropin** or **recombinant hGH.** Under appropriate conditions, cultures of engineered *E. coli* cells make almost limitless amounts of hGH which can then be isolated, purified, and used to treat hGH-deficient children. A child, treated with this recombinant hormone product, can grow as much as 5 inches in a single year.

hGH AFFECTS MORE THAN GROWTH The relatively large supply of hGH generated by transformed *E. coli* cells has provided a source of hormone sufficient to allow analysis of other beneficial effects of exogenous hGH. Research is currently under way to determine the possible benefits of recombinant hGH in other human pathologies. In general, hGH appears to hold promise in areas in which the body responds to extreme physical stress, including the healing of burns and other massive injuries, the prevention of calcium loss in the bones of the elderly, and the normal process of aging. In July 1990, scientists presented preliminary evidence suggesting that recombinant hGH may even reverse some of the effects of aging. Dr. Daniel Rudman and his colleagues at the Universities of Wisconsin and of California performed experiments in which administration of recombinant growth hormone resulted in an increase in lean body mass by almost nine percent, and a loss of nearly fifteen percent body fat. In addition, subjects in this study showed a thickening of the skin, which is associated with younger individuals, and an increase in growth-promoting hormones in the blood. Although these are preliminary results, it seems clear that recombinant hGH has not yet reached its full potential in therapeutic functioning.

INSULIN A hormone normally secreted by islet cells of the pancreas, serves to regulate the blood sugar level by aiding in the conversion of blood sugar to **glycogen**—a storage form of sugar. An individual lacking in insulin suffers from the disease known as **diabetes** which, when left untreated, results in extraordinarily elevated blood sugar levels eventually leading to diabetic coma and death. Diabetes can be controlled by the administration of exogenous insulin. Prior to 1982, most of the insulin used for this purpose was collected from the pancreases of cows and pigs. Although this animal insulin is similar enough to human insulin to provide relief from diabetic symptoms, it differs enough to provoke allergic reactions in some diabetic patients. Such patients did not obtain adequate relief from their symptoms using only animal insulin. These allergic reactions can only be avoided by using human insulin in place of animal insulin to treat the disease. The development of genetically engineered insulin completely avoids the problems associated with animal insulin and, in addition, avoids the problems of supply of the natural hormone when the livestock market fluctuates: *E. coli* cultures are not subject to such market value fluctuations!

Two chains, known as the **alpha chain** and the **beta chain,** compose the functional insulin molecule. The complexity of this molecule makes it necessary to use two types of transformed host cells in the cloning of human insulin. When insulin is produced by the techniques of genetic engineering, the DNA which encodes the alpha chain is used to transform one group of *E. coli* host cells, while another group of *E. coli* host cells is transformed with the DNA for the beta chain. The two different subunits, collected individually from the bacterial factories and chemically linked in the lab, produce a product identical to the naturally made human insulin molecule. The engineered product, originally developed by scientists at Genentech which is located in South San Francisco Bay area, is currently mass produced and marketed under the name of **Humulin** by Eli Lilly and Company. Doctors conservatively estimate that well over half of all diabetics who require injections of insulin are currently treated with Humulin.

Lymphokines Modulate the Immune Response

Lymphokines are a class of molecules that are naturally secreted by cells of the immune system. These molecules play an important role in the regulation of the immune response by acting as messengers between the various cellular players in immune function. Many different types of lymphokines are currently known, and each one plays a different specific and important role in the healthy functioning of the body.

INTERFERON One type of lymphokine, known as an **interferon,** was first discovered some thirty years ago when virologists noticed that cells infected with one type of virus were no longer susceptible to infection by a second type of virus. Scientists investigating this phenomenon isolated a molecule from infected cells that, when applied to healthy cells, appeared to interfere with subsequent viral infection of those cells. That molecule came to be called interferon. Since then, many interferon-type molecules have been discovered. Those molecules can be divided into three classes known as **alpha, beta,** and **gamma interferons.** Early analyses of the various types of interferons showed that not only did these molecules have a potent anti-viral effect, but they also actively inhibited the growth of cancer cells. It was obvious that further research on these remarkable molecules was warranted. Unfortunately, such research was delayed because, although interferon molecules are a normal product of normal cells, they are synthesized in extremely small quantities. This aspect of interferon production, coupled with the difficulty in extracting pure interferon molecules from body fluids and tissues, made it extremely difficult, if not impossible, to generate a supply of interferon molecules sufficient to allow research to proceed.

The advent of genetic engineering techniques made it possible for scientists to place interferon genes into *E. coli* host cells. Such transformed bacterial cells then synthesized these valuable molecules in almost unlimited amounts. The interferon preparations gained from these procedures made a great variety of research projects possible. The first application of an interferon molecule as a therapeutic substance was approved by the FDA in 1986 when alpha interferon was approved as a treatment for the rare cancer known as **hairy cell leukemia.** Interferons also decrease the rate of growth of a variety of other cancerous cells including multiple myeloma, Kaposi's sarcoma, malignant melanoma and some cancers of the kidney. For reasons which remain unknown, interferons are less effective against certain types of cancers including cancers of the breast, lung and colon. Research in the past few years shows that although interferons have some therapeutic value when applied to various types of cancers, they appear to be most effective when used in conjunction with other forms of cancer therapy including chemotherapy.

INTERLEUKINS ARE ALSO LYMPHOKINES Dr. Steven Rosenberg, chief surgeon at the National Cancer Institute of the National Institutes of Health, currently uses a type of lymphokine known as **interleukin-2,** or **IL-2,** in the fight against cancer. This lymphokine, normally secreted by activated T lymphocytes, serves to stimulate more rapid T cell proliferation. Dr. Rosenberg isolates cells from the immune system of a cancer patient and grows them in growth medium supplemented with IL-2, hoping that the immune cells will be activated by the lymphokine. Dr. Rosenberg then returns the cells, now known as **lymphokine activated killer cells,** or **LAK** cells, to the patient along with additional doses of genetically engineered IL-2. Although Dr. Rosenberg currently achieves a complete remission in only one patient out of ten, due to the experimental nature of this therapy only patients who have exhausted all other treatment options are currently being treated with the LAK/IL-2 regime. These are patients for whom nothing else has worked, and who have no other hope. While most patients receiving this experimental regime do not experience complete remission, many of them demonstrate a remarkable shrinkage of cancerous tumor masses.

Dr. Rosenberg is currently working on two modifications of his original LAK/IL-2 treatment. In one modification Dr. Rosenberg collects cells of the immune system directly from the tumor mass, rather than from the general blood supply. These cells, called **tumor infiltrating lymphocytes,** or **TILs,** are thought to be present in the tumor because they possess an inherent ability to act against cancer cells. The TILs, treated with IL-2 prior to being reinserted, are transfused into the body of the cancer patient along with additional genetically engineered IL-2. Although this modification appeared to show great improvements over the original protocol when it was tested in laboratory animals, in preliminary trials it has so far failed to live up to expectations when used in human patients: Although the LAK/IL-2 or the TIL/IL-2 combinations are not panaceas, they do hold some promise in the treatment of cancer.

TUMOR NECROSIS FACTOR IS A LYMPHOKINE The second type of modification which Dr. Rosenberg is investigating involves the world of **gene therapy.** This term, which has a slightly different meaning from the gene replacement therapy discussed above, refers to the insertion of extra or therapeutic genes so that their gene products act to modulate a particular phenotype. Thus, gene therapy differs from gene replacement therapy in that it involves the introduction of extra genetic material rather the simple replacement of a mutant gene.

Tumor necrosis factor, or **TNF,** a lymphokine which both kills tumor cells, and causes a generalized wasting syndrome called **cachexia** when administered to laboratory animals or to humans. Therefore, scientists would like to deliver TNF directly to cancer cells without having to administer it through the general blood supply or through non-cancerous tissues. Dr. Rosenberg is collaborating with Dr. W. French Anderson in an attempt to introduce the gene which encodes TNF into certain cells of the immune system. Drs. Rosenberg and Anderson believe that if the TNF can be delivered directly to the cancerous tumor cells by the transformed immune cells, the cachexia might be averted while the ability to kill cancer cells is retained.

The first of such experiments, carried out in January 1991, involved two cancer patients, each suffering from metastatic melanoma, a lethal skin cancer. These patients, a twenty-nine-year-old woman and a forty-two-year-old man, were each infused with TILs which had been transformed with the gene which encodes TNF. It is hoped that the negative side effects generated by systemic administration of TNF will be avoided since the TIL cells will, ideally, quickly home in on cancer cells while leaving healthy cells alone. In this way it is predicted that the anti-cancer effect of TNF can be restricted to the involved site or sites, rather than bathing all of the tissues of the body in TNF. Once again, only time will tell us of the efficacy of this approach.

Cells as Drug Delivery Systems: Magic Bullets

Another area of biotechnology searches for a **magic bullet,** a cell which could be loaded with an appropriate treatment molecule and, which when injected into a patient, would migrate through the circulatory system directly to the diseased site. The collaborative efforts of Drs. Rosenberg and Anderson described above fall under the general category of magic bullet research because they use cells as vehicles to deliver drugs to the site where they are most needed. Other groups of scientists are also specially designing cells with the ability to deliver a specific drug or therapeutic agent to a known site. Perhaps someday not only will anti-cancer drugs be delivered to a tumor but the progression of Alzheimer's disease will be slowed or even halted by a cell which goes directly to the brain tissue and delivers an enzyme to prevent or remove the formation of amyloid plaques. Maybe the deadly progression of cystic fibrosis will stop when a cell can go directly to the lung tissue and deliver an enzyme to break down the thick gelatinous mucous which characterizes this disease. As exciting as these prospects are, we need much more research to make them into reality!

Enzymes as Therapeutic Agents

Tissue plasminogen activator, or **tPA,** a protein produced by the cells of the body, is involved in the dissolution of unwanted or unnecessary blood clots. Like interferons, tPA is produced by normal cells of the body, but in quantities so small that difficulty arises in isolating sufficient amounts for either research or therapeutic use. Scientists can clone the human tPA gene into bacterial cells and

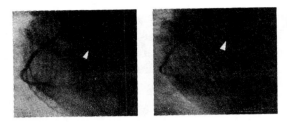

Figure 11.5
Tissue Plasminogen Activator acts to dissolve blood clots designated by arrows thereby allowing normal blood flow to resume.

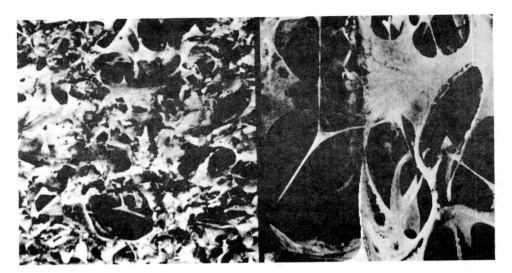

Figure 11.6
Emphysema damages the delicate structure of lung tissue, thus rendering its victims unable to breathe effectively. This scanning electron micrograph shows both normal lung tissue (left) and lung tissue destroyed by emphysema (right).

generate almost limitless amounts of this enzyme. When the victim of a heart attack, brought on by a blood clot which blocks a cardiac vessel, receives tPA via intravenous injection, the tPA travels directly to the offending blood clot and eats through it much as drain cleaner works on a blocked drain (Figure 11.5). Although other "cardiac-drano" type drugs are available, most obviously streptokinase and urokinase, these drugs must be administered directly to the blocked vessel and do not work as quickly as tPA.

Alpha-1 Antitrypsin

Emphysema, a disease in which the delicate balloon-like structure of the alveoli of the lungs is eaten away, causes the victim to find breathing to be a painful and ineffective process (Figure 11.6). Although smoking appears to play a significant role in this progressive and lethal disease, a significant number of emphysema patients possess a mutant form of the gene for the enzyme known as

alpha-1 antitrypsin or **AAT.** This enzyme functions in the normal individual by protecting the lung tissue from destruction by the naturally present enzyme **neutrophil elastase.** Dr. Ronald Crystal of the National Heart, Lung, and Blood Institute envisions a cure for emphysema whereby the gene for AAT, introduced into a viral cloning vector, would in turn be introduced into a T lymphocyte. An aerosol spray would then deliver such engineered cells directly to the lungs where they would manufacture a protective enzymatic coat for the vulnerable lung tissue. Preliminary experimental results presented in 1989 showed that aerosolized AAT does indeed protect lung tissue from the ravages of neutrophil elastase and that lymphocytes transfected with an AAT-containing retrovirus did, indeed, lead to the expression of human AAT when introduced into mice. Attempts at this type of gene therapy in human patients awaits further research and eventual approval by the appropriate government committees. If this therapy proves effective, similar therapies may be developed against other pulmonary diseases, including cystic fibrosis in which the lungs become clogged with abnormally thick and gelatinous mucous. Scientists would like to introduce a cell containing a gene which synthesizes an enzyme able to break down the mucous in an attempt to alleviate the symptoms of this disease.

TRADITIONAL VACCINES MADE FROM WHOLE VIRUS

In 1796 an epidemic of **smallpox,** an often fatal viral disease, swept through the English countryside killing one quarter of the population. A physician named Edward Jenner noticed that only two groups of people escaped infection: those who had previously contracted smallpox and survived the infection, and milkmaids who had caught a pox-like infection from the cows they milked. This infection, called **cowpox,** took a much milder form than the deadly smallpox. Dr. Jenner hypothesized that infection by cowpox could somehow confer protection against infection from smallpox. To test this hypothesis Dr. Jenner isolated material containing the causative virus from the pox of people infected with cowpox and injected healthy people with this material. Although this treatment caused a mild illness in the recipients, it appeared to confer protection against smallpox on the treated individuals. This treatment came to be known as **vaccination,** a word derived from the Latin *vacca* meaning "**cow.**"

In modern days the term "vaccination" has taken on a more specific meaning: the activation of the immune response by presentation of an **attenuated** (weakened) or killed **pathogen,** or disease-causing microorganism. The immune activation which results from such a procedure allows a quick response to any subsequent future exposures to the same pathogen, even in its strong and unweakened state.

Pathogens Play Two Roles in Disease

Pathogens have two functions within our bodies; one function causes illness while the other function results in the activation of the immune response. The illness-causing function is the end-result of activities directed by the nucleic acids of the pathogen. The immune-response function results from the presence of certain molecules—called **antigens**—on the surface of the pathogen which are recognized by the body as being foreign. When the immune system is activated it does two things. First, it works to defeat the pathogen that caused its activation. Second, it produces special cells called **memory B cells.** These memory cells remember the presence of the pathogen so that they can quickly mount an effective immune response any time that the same pathogen is reencountered, even many years in the future. This quick response protects a person from multiple infections by the same virus: for example, if you have a functioning immune system and you get viral chicken pox once, you will never get it again. However, if you get chicken pox during a time when your immune system is not fully functional, such as when you were a young baby, your body cannot produce memory B cells. In this unique circumstance, your immune system lacks the memory necessary to mount the quick immune response needed to avoid illness upon a subsequent exposure to the virus.

Jenner's successful use of cowpox vaccines to protect against smallpox results from the similarities of the molecules on the surface of the two types of viruses. The cowpox molecules fooled the body's immune system into thinking that it had been activated by the smallpox molecules. In this way, when the vaccinated individual was later exposed to smallpox, his or her immune system immediately leaped into action and defeated the deadly virus before it got established in the body.

Traditional Vaccines Present Some Problems

Until 1981 all vaccines were made using a modification of the same method that was used by Edward Jenner: pathogens were attenuated or killed with the intention of disabling the disease-causing function of the pathogen while the immune-activating molecules remained unharmed. These modified pathogens, when injected into a person, induce immunity to the unmodified pathogen without causing the disease itself.

Unfortunately, this so-called **whole virus** vaccine also comes with its problems. In some cases attenuated pathogens revert back to their virulent or unweakened state, thus causing disease in the vaccinated person rather than preventing it. Similarly some pathogen particles could possibly escape attenuation or being killed. In addition, some vaccines prepared in this way contain impurities stemming from the methods of production which usually rely on various animal species. These impurities, often expensive, difficult, or even impossible to remove, can result in serious adverse reactions if they remain in the preparation. Despite the potential problems, whole virus vaccines have largely eradicated a variety of diseases, including smallpox and polio, and have caused the dramatic decline of yellow fever, pertussis, measles, mumps, rubella, tetanus, and diphtheria.

Recombinant Vaccines Don't Cause Disease

Genetic engineering techniques alleviate potential vaccine-induced illnesses by allowing scientists to isolate the immune activation function of pathogens and separate it from the disease-causing functions. So-called **recombinant vaccines,** or **subunit vaccines,** rely on the ability of the scientist to isolate the gene or genes responsible for encoding the cell surface molecules which trigger the immune-activation function. These genes can then be removed from the pathogen's genome by endonuclease action and inserted into a cloning vehicle which is, in turn, used to transform a bacterial or yeast host cell. The host cell then synthesizes a supply of the antigenic molecule which can then be isolated, purified, and made into a vaccine. When that vaccine is injected into a person, that individual mounts an immune response against that antigen. The immune response includes the synthesis of appropriate memory B cells. If the immunized individual is ever exposed to a pathogen which carries that particular antigen on its cell surface, an immune response quickly eliminates the entire pathogen. Since the recombinant vaccine only contains the immune-activating subunit of the pathogen, the disease state cannot be induced. For this reason recombinant vaccines are sometimes called **subunit vaccines**—they only contain a subunit of the original pathogen.

The first vaccine to be produced by these techniques, completed in 1981 and designed for use in animals, has largely eliminated **hoof-and-mouth disease.** The first recombinant vaccine designed for human use took almost fifteen years to develop and test and was finally approved by the FDA in 1986. This vaccine combats hepatitis B.

Some Vaccines Are Made from Vaccinia Virus

Vaccinia virus, also known as the cowpox virus, is relatively harmless in its effects on humans, and as described above, has been used as the basis of the smallpox vaccine. The vaccinia virus is also the basis of a type of vaccine which combines features of both a whole virus vaccine and a recombinant subunit vaccine.

In order to make such a vaccine scientists use recombinant DNA techniques to genetically alter the genome of the vaccinia virus by the insertion of foreign genes which encode pathogenic antigens. These genes take the place of the endogenous vaccinia genes which, if present, would induce immunity to smallpox. When a person is immunized with a vaccinia preparation which has been modified in such a fashion, the spliced-in foreign gene enters the cells of the vaccine recipient and directs the synthesis of the foreign antigenic protein. This protein then activates the immune system in a way which resembles other types of vaccines. The vaccinia virus can be modified in such a way that a single preparation confers protection against a series of diseases. Scientists hope someday to modify a vaccinia preparation so that it protects against as many as twelve different diseases. Such a vaccine would change the face of health care in many third-world countries.

Vaccinia Used to Combat Rabies

Rabies, a disease transmitted in saliva from the bites of infected animals, often dogs or raccoons, is invariably fatal if left untreated. Once the rabies virus enters the body it travels to the brain where it causes extensive tissue damage leading to convulsions, paralysis, insanity, and, eventually, death. Two different types of rabies vaccines are currently available. One vaccine, used to treat human victims of a rabid bite, must be administered soon after the bite occurs. This vaccine, which is almost always effective, requires several injections and costs close to $1,000 per patient. However, additional treatment is painful, and sometimes the patient cannot receive the injections soon enough to ward off development of the disease. The other type of vaccine is used to treat animals. While this vaccine is effective, it must be injected into each and every animal on an individual basis, making it impossible to vaccinate a sufficient segment of a wild animal population to eradicate the disease.

Scientists have employed the techniques of genetic engineering to synthesize still another type of anti-rabies vaccine for animals. This vaccine uses a small piece of the rabies virus genome which is inserted into the genome of the vaccinia virus. The new recombinant virus, originally developed in 1984 by scientists at the Wistar Institute in Philadelphia, cannot cause rabies in those animals that receive it. Instead, the recombinant vaccinia vaccine contains only those rabies genes that encode antigenic molecules and therefore has the sole function of activating the immune response against rabies infection. Perhaps most importantly, this new vaccine can be administered orally, thus allowing a phenomenal increase in the number of animals which are potential recipients of the vaccine. Such an increase will cut down on the number of rabies-infected animals, and therefore on the number of humans who contract rabies from the bite of an infected animal. The necessary testing of this vaccine currently awaits approval by various government and state regulatory agencies.

Vaccines Are Specific in Their Functioning

All vaccines, either whole virus or subunit, function by inducing an immune response against a specific subset of molecules appearing on the surface of the pathogen. For this reason, a vaccine prepared against one influenza strain, for example, rarely works against a different influenza strain. Consequently, health workers often advise the very young, the very old, and the chronically ill to receive a flu shot each and every year. Although the flu comes around with great regularity, each outbreak is generally caused by a new and different influenza virus, thus requiring new vaccines for each outbreak of the illness. The same is true of vaccines which are active against other viral illnesses. For example, researchers had to develop vaccines against three different strains of polio virus in order to eradicate this single disease.

AIDS AND THE SEARCH FOR A VACCINE

Human Immunodeficiency Virus, or **HIV** (Figure 11.7), is a member of a class of viruses known as **retroviruses.** These viruses differ from the majority of cell types in that they contain RNA as their sole genetic material rather than the more common DNA. When the HIV retrovirus infects a cell it injects its RNA and, along with that genetic material, an enzyme known as **reverse transcriptase.** Once inside the host cell, the reverse transcriptase directs synthesis of a DNA molecule complementary to the viral RNA. This virally derived DNA then goes on to lead the infective assault on the host cell and results in the condition that we now call **Acquired Immunodeficiency Syndrome** or **AIDS.**

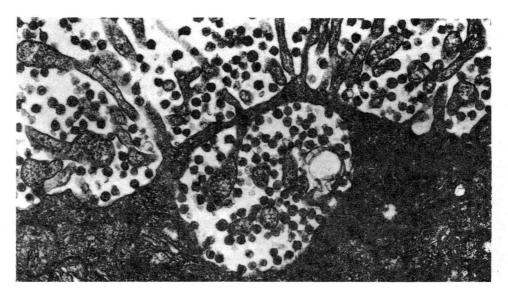

Figure 11.7
This micrograph shows examples of the Human Immunodeficiency Virus. HIV, is a retrovirus, 1/10,000th of a mm in diameter. Inside the virus particle is the RNA which holds the genetic information. The RNA is combined with a variety of proteins to form a bullet shaped core within the virus particle. The surface of the particle holds knob-like structures which consist of three different proteins and which function in binding to a cell prior to infection. Although research continues, there is currently no known cure for HIV infection.

A person infected with HIV follows a frighteningly predictable course of symptoms. Immediately after exposure to HIV, a person experiences a bout of flu-like symptoms. While this "flu" may appear unusually severe, it eventually disappears. This doesn't mean, however, that the virus has been eliminated from the body. Instead, HIV attaches to special cells of the immune system called **helper T cells** or, sometimes, **T-4 helper cells.** This attachment is made when a molecule on the surface of HIV called **gp120** attaches to a receptor on the helper T cell known as **CD4.** Soon after this attachment HIV enters the helper T cell. HIV can also infect other types of cells including **macrophages,** which function to remove debris from body tissues, and various cells of the brain. In many cases the HIV particles contained within cells of the body become dormant. This period of dormancy, known as the **latent period,** lasts between two and ten years, with most people experiencing a latent period somewhere between these two extremes. During the latent period, the presence of the virus slowly destroys helper T cells. As the number of helper T cells diminishes, immune function also decreases. During this time the patient begins to exhibit an array of physical symptoms which often include **chronic lymphadenopathy,** or swollen lymph glands, and generalized fatigue. When the number of T cells drops below a certain level, the disease enters a more active phase. Eventually most AIDS patients begin to develop diseases which are easily handled by a healthy immune system and, for that reason, are seldom seen in the uninfected population. As a result AIDS patients typically display a variety of pathological conditions including **Kaposi's sarcoma,** a tumor of the lining of the blood vessels, and a protozoal pneumonia known as **pneumocystis carinii pneumonia.** With the continued destruction of the helper T cells, the HIV-infected individual eventually loses all immune function.

Interestingly, as doctors become better able to combat typical AIDS-associated diseases, patients live longer but succumb to a variety of new illnesses. For example, in the spring of 1990 doctors at the National Cancer Institute announced a dramatic increase in the number of AIDS patients developing cancer of the lymph system. To put it in perspective, data for San Francisco show that cases of Kaposi's sarcoma increased at a rate of two percent per year while cases of lymphoma increased by forty-eight percent.

Although doctors have made great strides forward in treating the symptoms of AIDS, including the various associated infections, pneumonias, and cancers, they have yet to find a way to restore an immune system ravaged by the disease. Many people suggest that the appropriate way to proceed at this point is to develop a vaccine against the infection.

Unfortunately, HIV is not amenable to vaccine production for a variety of reasons:

1. HIV is an extremely complex virus. It can hide in the interior of infected cells so that it leaves no trace of viral antigens on the cell surface. As a result, it tricks the immune system into behaving as though no viral intruder were present. In addition the genetically variable HIV has the ability to modify many of the antigenic molecules on its surface over time.

2. A good animal model has not been found for human HIV infection. As a result it is nearly impossible to achieve adequate testing of experimental preparations and research is unavoidably slowed.

3. A great many ethical considerations are involved with the necessary human testing of any potential AIDS vaccines.

4. HIV, a retrovirus and characteristic of such viruses, has the ability to insert pieces of its own genome into the genome of the infected host. Such a scenario renders the viral genetic material invisible to the host immune system.

Despite these problems a variety of vaccines have been tested, including one prepared by scientists at Genentech, Inc. in South San Francisco, California. The Genentech research team prepared a subunit recombinant vaccine involving the gp120 component of the outer coat of the AIDS virus. Scientists tested this preparation using three chimpanzees, two of which received the experimental vaccine. All three of the animals then received injections of the same strain of AIDS virus as was used in the preparation of the vaccine. The unvaccinated animal developed AIDS infection seven weeks after immunization while the vaccinated animals had yet to show signs of infection six months after the experiment. While this vaccine appears promising, questions remain. Chimpanzees do not provide an optimum model of human AIDS since they do not develop a human-like pattern of HIV infection. In addition, this vaccine was made against a single strain of the AIDS virus. Multiple strains of HIV likely exist, and it is unclear whether this vaccine will provide protection against additional strains or whether multiple vaccines will be required.

Four other types of AIDS vaccines are currently being tested in the United States, and another type which consists of a mixture of vaccines, all directed against a different segment of the AIDS virus, is being tested in France.

DNA PROBES AND MEDICINE

One of the most interesting ways in which recombinant DNA and medicine intersect is when genetically engineered DNA probes aid in the diagnosis of human diseases. As discussed above and in Chapter Five, one type of probe is a radioactively labelled DNA fragment which is complementary to a specific gene or gene segment. Probes, used to analyze the contents of gene libraries, thus identify target genes. In a similar fashion, appropriate probes can be used to analyze the human genome and identify mutant or abnormal genes. Prenatal medicine currently uses probes to determine the presence of the genes for both Tay-Sachs disease and cystic fibrosis, to name just two. This type of analysis takes advantage of the ability of the physician to withdraw some of the **amniotic fluid**—the fluid which surrounds and cushions a growing fetus—from the uterus of a pregnant woman. Fetal cells, isolated by centrifugation from this fluid, grow under laboratory conditions. The DNA is then collected from these cells, denatured, and exposed to the labelled probe. Binding of the probe to the fetal DNA indicates the presence of the target gene. In other words, if the probe used complements the gene that accounts for the development of Tay-Sachs Disease and binds to DNA in the fetal cells, it can be concluded that the fetal cells contain the Tay-Sachs gene.

Genetic probes are also beginning to play a role in the rather controversial area of **predictive medicine.** Predictive medicine attempts to identify persons at risk for the development of certain diseases prior to the onset of symptoms. This feat requires that the appearance of something in the genome—either a mutant gene or a RFLP—be coordinated with the future onset of a disease. For example, scientists have determined that a particular RFLP appears to be coordinated in its expression with the development of Huntington's disease. Since Huntington's disease does not typically manifest itself until the victim reaches adulthood, often forty years of age or more, it carries with it an aura of uncertainty. A person with an affected family member may live for a great many years without knowing whether they, too, will eventually develop the disease. By analyzing the genomes of potential victims for the presence of the Huntington's-related RFLP, doctors can now counsel individuals regarding their chances of developing the disease. Such information could well affect the life choices of that individual—should I marry? become a parent? etc.

Emphysema is another disease which is amenable to probe analysis. As discussed previously, emphysema patients often have abnormally low levels of an enzyme called alpha-1 antitrypsin due to a single genetic mutation, which is detectable in individuals prior to the onset of the disease. Thus, genome analysis with a probe specific for the mutant alpha-1 antitrypsin gene may identify individuals who are at a higher risk for the development of emphysema. Such individuals, armed with this type of information, may choose to modify their lifestyles accordingly.

REVIEW AND SUMMARY

1. Almost four thousand genetic diseases which affect humans have been described. Three hundred of those are **metabolic disorders** which involve the absence of a particular enzyme. Metabolic diseases involve two levels of defect: the gene itself and the gene product.

2. Various procedures have been used in the attempt to assign genes to their chromosomal map positions. These procedures include mapping with **somatic cell hybrid clone panels,** analysis of **FACS-sorted chromosomes,** and association of unknown genes with **markers** of known position. Markers, which can include RFLPs as well as genes, provide landmarks for the analysis of chromosomal positions. Approximately one thousand genetic markers have been defined which span the human genome at intervals of approximately ten million base pairs. Areas located between two markers can be accessed by the technique of **chromosome walking.**

3. The **Human Genome Mapping Project,** a long term project, will identify many new markers, assign all human genes to chromosomal map positions, and determine the base sequence of the entire human genome. This information could aid in new treatments, therapies, and diagnostic techniques for human diseases, as well as provide insight into various questions of basic scientific research.

4. **Gene replacement therapy** is the replacement of a faulty gene with a normal gene in an attempt to correct a genetic illness. For this technique to be completely successful, the replacement gene must be inserted into the appropriate stem cells so that it replicates and passes onto all descendants of that stem cell as would occur in normal development. In addition, the inserted replacement gene must be placed in a chromosomal position which allows it to be under correct control of expression and does not damage the function of another gene. If the new gene is inserted into a germ cell, it will be passed onto future generations. If the gene is inserted into a somatic cell, future generations will remain unaffected.

5. **Gene therapy** is the insertion of an extra gene with the intention that the gene product will play a therapeutic role. Some scientists are interested in the development of a **"magic bullet."** This cell would contain a therapeutic gene and would travel directly to the site of illness. Once at the target site, the gene product would play a role in the alleviation of the disease state.

6. The techniques of biotechnology have made it possible to produce a wide variety of **therapeutic molecules** which were previously available only in short supply. These molecules include **hormones** such as human growth hormone and insulin, **lymphokines** such as interferons, tumor necrosis factor, and interleukins, and **enzymes** such as alpha-1 antitrypsin and tissue plasminogen activator.

7. **Vaccines** have traditionally been made by attenuating or killing virus particles and using the resulting material to induce an immune response in vaccinated individuals. Although such vaccines have enjoyed a remarkable success and are responsible for the near eradication of polio and other diseases, whole virus vaccines have certain problems. Specifically, some virus particles can escape the weakening or killing process used in the preparation of the vaccine. The vaccine which results from this situation has the potential of inducing the viral illness in the individual that receives the vaccine.

8. **Recombinant,** or **subunit, vaccines** avoid this difficulty. By cloning the gene responsible for encoding the immune-activating viral surface antigens and discarding the rest of the genome, scientists can create vaccines without the ability to induce viral illness. These vaccines are composed of only the viral antigens that are produced by host cells which have been transformed with the appropriate viral gene or genes.

9. Some vaccines are made by modifying the genome of the relatively harmless **vaccinia virus** so that it includes the antigen genes of other more pathogenic viruses. Such a vaccine can be easily administered and may even be engineered to confer protection against more than one disease.

10. The nature of the AIDS virus makes it extremely difficult for scientists to develop a vaccine. Not only is the AIDS virus genetically unstable, but it can hide in the interior of host cells. In addition, researchers have yet to discover a good animal model of this disease in which to test potential vaccines, and moral difficulties hinder any proposed testing of an AIDS vaccine on healthy human volunteers. Despite these problems, a potential vaccine prepared against the gp120 component of the HIV coat has conferred protection on two chimpanzees in preliminary experiments.

11. **Probes** which are specific to various genes and DNA fragments have made a wide variety of prenatal diagnoses possible. In addition, genetic probes helps predict the possible future onset of certain diseases, including Huntington's disease and emphysema.

Applications of Biotechnology: Agriculture

CHAPTER 12

Many of us are "weekend farmers." Whether you live in the city and plant in a window box, or in the suburbs and work in your backyard, at one time or another you have probably tried your hand at vegetable gardening (Figure 12.1). Perhaps you yearned for that perfect tomato, or maybe you wanted to decrease your out-of-pocket expenses for vegetables. Whatever the reason, most people attempt tiny-scale farming at some time in their lives.

If you are one of these people you most likely encountered some common garden pests. Perhaps snails ate your lettuce or worms ate your corn. If that happened, you faced a few options: you could either write off your gardening experience, or you could use some type of pest control. If you battled the bugs, or perhaps you choose a toxic pesticide and then worry about possible toxicity to you and your family as well as to the environment. Or perhaps you chose an organic preparation, but it was more difficult and expensive to use than its synthetic counterparts. Sometime during your battle you probably wished that the vegetable plants came with a built-in pest repellent. It would be easier, would cost less, and would probably be less toxic all around.

This wish is something that genetic engineers and biotechnologists have been thinking about for many years. This chapter discusses some of the many levels of involvement of biotechnology and agriculture.

OVERVIEW The advances of biotechnology play an important role in many areas of agricultural development, including animal husbandry, crop nutrition and resistance to pests, and food production. This chapter briefly examines some of those topics and answers the following questions:

1. How might plants produce their own fertilizers?
2. Can some bacterial infections really be good for a plant? Why or why not?

Figure 12.1
Most people have tried their hand at gardening at one time or another.

3. Can we get complete protein nutrition from plant sources? Can plant proteins be modified?

4. What effect does superovulation have on animal fertility?

CLONING OF PLANT CELLS AND MANIPULATION OF PLANT GENES

Plant cells exhibit a variety of characteristics which distinguish them from animal cells. These characteristics include the presence of a large central vacuole and a cell wall and the absence of the centioles which play a role in mitosis, meiosis, and cell division. Along with these physical differences another factor distinguishes plant cells from animal cells and is of great significance to the scientist interested in biotechnology: many varieties of full-grown adult plants can regenerate from single modified plant cells called **protoplasts**—plant cells whose cell walls have been removed by enzymatic digestion. More specifically, when some species of plant cells are subjected to removal of the cell wall by enzymatic treatment, they respond by synthesizing a new cell wall and eventually undergoing a series of cell divisions and developmental processes that result in the formation of a new adult plant. That adult plant can be said to have been cloned from a single cell of a parent plant. Relatively easily clonable plants include carrots, tomatoes, potatoes, petunias, and cabbage, to name but a few. This important ability to grow a whole plant from a single cell means that researchers can engage in the genetic manipulation of just a single cell, allow that cell to develop into a complete mature plant, and examine the whole spectrum of physical and growth effects of the original manipulations. Such a process is far more straightforward than the parallel process in animal cells. Single animal cells cannot be cloned into full grown adults, thus rendering an examination of most of the effects of genetic manipulation on animal cells far more difficult.

A CLONING VECTOR THAT WORKS WITH PLANT CELLS

Not all aspects of the genetic manipulation of plant cells are so easily accomplished. Not only do plants usually have a great deal of chromosomal material and grow relatively slowly as compared with single cells grown in the laboratory, but few cloning vectors can successfully function in plant cells. While researchers working with animal cells can choose among a wide variety of cloning vectors to find just the right one, plant cell researchers are currently limited to just a few basic types of vectors. Perhaps the most commonly used is the **Ti plasmid** or **tumor inducing plasmid.** This plasmid, found in cells of the bacterium known as *Agrobacterium tumefaciens,* normally lives in soil, and has the ability to infect plants and cause a **crown gall,** or tumorous lump, to form at the site of infection. The tumor-inducing capacity of this bacterium results from the presence, within infecting cells, of the Ti plasmid. The Ti plasmid itself, a large, circular, double-stranded DNA molecule, can replicate independently of the *A. tumefaciens* genome. When these bacteria infect a plant cell, a 30,000 base-pair segment of the Ti plasmid—called **T DNA**—separates from the plasmid and incorporates into the host cell genome. This aspect of Ti plasmid function has made it useful as a plant cloning vector.

The Ti plasmid can be used to shuttle exogenous genes into host plant cells. This type of gene transfer requires two steps (Figure 12.2): first, the endogenous tumor-causing genes of the T DNA must be inactivated and second, foreign genes must be inserted into that region of the Ti plasmid. The resulting recombinant plasmid, carrying up to approximately 40,000 base pairs of inserted DNA and including the appropriate plant regulatory sequences, can then be placed back into the *A. tumefaciens* cell; that cell can be introduced into plant cell protoplasts either by the process of infection or by direct insertion. Once in the protoplast, the foreign DNA, consisting of both T DNA and the inserted gene, incorporates into the host plant genome. The engineered protoplast—containing the recombinant T DNA—regenerates into a whole plant, each cell of which contains the inserted gene. Once a plant incorporates the T DNA with its inserted gene, it passes on to future generations of the plant with a normal pattern of Mendelian inheritance.

One of the earliest experiments that involved the transport of a foreign gene by the Ti plasmid took place when scientists inserted a gene isolated from a bean plant into a host tobacco plant. Although this experiment served no commercially useful purpose, it successfully established the ability of the Ti plasmid to carry genes into plant host cells where they could be incorporated and expressed.

A. Tumefaciens Infects a Limited Variety of Plant Types

The fact that only certain types of plants were naturally susceptible to infection with the host bacterial organism initially limited the usefulness of Ti plasmid as a cloning vector. In nature, *A. tumefaciens* infects only **dicotyledons** or **dicots**—plants with two embryonic leaves. Dicotyledenous plants, divided into approximately 170,000 different species, include such plants as roses, apples, soybeans,

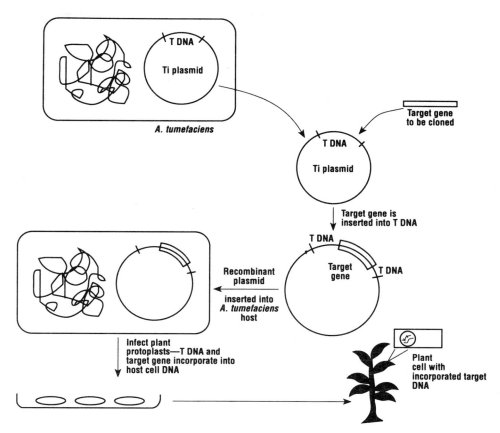

Figure 12.2
As described in the text, this figure illustrates the procedures used in Ti-plasmid cloning.

potatoes, pears, and tobacco. Unfortunately, many important crop plants including corn, rice, and wheat, are **monocotyledons**—plants with only one embryonic leaf—and thus could not be easily transfected using this bacterium.

Overcoming the Limited Range of A. Tumefaciens Infection

Research efforts in the past few years have reduced this limitation somewhat. Scientists discovered that using the processes of **microinjection, electroporation** and **particle bombardment,** naked DNA molecules can be introduced into plant cell types which are not susceptible to A. tumefaciens transfection. As discussed previously, microinjection involves the direct injection of material into a host cell using a finely drawn micropipette needle. Electroporation uses brief pulses of high voltage electricity to induce the formation of transient pores in the membrane of the host cell. Such pores appear to act as passageways through which the naked DNA can enter the host cell. Particle bombardment actually shoots DNA-coated microscopic pellets through a plant cell wall. These developments, important in the commercial usefulness of plant genetic engineering, render the valuable food crops of corn, rice, and wheat susceptible to a variety of manipulations by the techniques of recombinant DNA and biotechnology.

RESISTANCE TO HERBICIDES

Herbicides are chemical agents that, when applied to various types of plants, result in the death of the plant. Herbicides are used primarily to kill non-desirable plants that compete with a desired plant for space, water, and nutrients. Problems arise when the desired plants are not entirely resistant to the effects of the herbicide.

A number of research groups carried out experiments in which genes conferring resistance to the widely-used herbicide **glyphosate** were inserted into a variety of crop plants including corn and tobacco. Glyphosate, commonly marketed under the trade name of Roundup or Kleenup, is an effective herbicide: it is extremely potent, yet it has few known toxic effects on animals and has a short half-life in the environment. It would be useful if crop plants could be infected with a bacterial cell carrying a Ti plasmid which, in turn, carried the gene for glyphosate resistance. If the inserted genes were actively transcribed and translated, they would confer resistance to glyphosate on their host plants. In such a situation, entire fields could be sprayed to eradicate undesirable plants and weeds, which compete with valuable crops for materials necessary for growth, while leaving the crop plants themselves unaffected by the herbicide.

RESISTANCE TO PESTS: "NATURAL" PESTICIDES

Pesticides and **insecticides** are chemical agents that kill various types of insects and other pests while leaving plants and most animals relatively unharmed. Unfortunately, many pesticides are non-selective: they kill insects which are beneficial to plants and the environment along with those which are not. In addition, pesticides can be toxic to humans and other non-target animals and they contaminate soil and water supplies. This environmental contamination may then lead to a second level of adverse effects on both plant and animal life. Furthermore, prolonged use of insecticides can lead to the development of insect strains which are no longer susceptible to the toxic effects of those chemicals. Such a development has already resulted in the appearance of DDT resistant mosquitos—the chemical DDT was most commonly employed in the battle against malaria-carrying mosquitos. Because of these problems, a major goal of biotechnologists has been to provide plants with a natural endogenous resistance to a variety of pests through the techniques of genetic engineering.

Bacillus thuringiensis, a bacterium, possesses endogenous insecticidal activity directed against a variety of insects including mosquitoes. To exploit this lethal characteristic, *B. thuringiensis* spores are combined with water to form a mixture which can be sprayed over an insect-infested area. When insect larvae ingest the bacterial spores, the spores release a chemical toxin which, in turn, causes the death of the insect larvae. This treatment, often used in the control of both mosquitoes and gypsy moths, effectively causes a decrease in these pest populations within a given area. However, the insecticidal effects are transient due to the limited field survival of the spores, so long-term insecticidal activity requires repeated applications of spores to the treated area.

Genetic engineers attempt to circumvent the need for repeated spore treatments with the techniques of recombinant DNA technology. In one line of research scientists isolated the genes which encode the *B. thuringiensis* toxin and inserted those genes into other organisms which are better suited for survival in the field. Examples of organisms which have been transformed in this manner include *E. coli* and the bacterium **Pseudomonas fluorescens.** *P. fluorescens* lives on the roots of many different types of plants, including corn. Tests show that when *P. fluorescens,* transformed with a cloning vector containing the genes for the *B. thuringiensis* toxin, is sprayed on corn plants, the transformed bacterial cells colonize the root area. As a result, the *B. thuringiensis* toxin is synthesized at the site of the plant itself, thus endowing the corn plant with certain endogenous insecticidal abilities.

A second line of research involved linking the *B. thuringiensis* toxin gene to a constitutively expressed promoter and introducing it, via the Ti plasmid, directly into the cells of the plant which is to be protected. One such study, which took place in Belgium, involved the insertion of the toxin gene into cells of the tobacco plant. Analysis of tissues of the mature transformed plants showed that cells of the tobacco plant indeed synthesized *B. thuringiensis* toxin. When the treated plants were subjected to infection with tobacco hornworm larvae, they survived with little, if any, damage. In contrast, control tobacco plants, which were not transformed with the engineered Ti plasmid, all died within approximately two weeks. The inserted gene clearly provided considerable protection to the infected tobacco plants. As with other research of this type, it remains to be seen if these results will provide useful crop protection in uncontrolled field situations.

Similar studies make use of the endogenous insecticidal activities of a type of virus known as **baculovirus.** Like *B. thuringiensis,* baculovirus particles can be combined with water and then sprayed onto foliage. When damage-causing insects eat the virus-treated foliage they are subject to a viral infection which leads to the death of the insect. Large scale field applications of both *B. thuringiensis-* and baculovirus-derived insecticides await further study and approval by appropriate governmental regulatory agencies.

NITROGEN FIXATION

All living things, including plants, require a source of nitrogen to synthesize essential carbon-containing molecules, such as amino acids, nucleic acids, proteins, and vitamins. Despite the fact that the atmosphere around us consists of approximately eighty percent nitrogen, this requirement presents no small problem: nitrogen in its atmospheric form is not usable by animals, plants, fungi, or the vast majority of bacterial strains. Before it can be used in essential growth processes, atmospheric nitrogen must be converted into appropriate nitrogen-containing compounds, usually by the addition of either hydrogen or oxygen. These modified nitrogen compounds can, in turn, be used in various biosynthetic processes. This process of conversion—called **nitrogen fixation**—is carried out only by a limited number of bacterial species, which then share fixed nitrogen with nitrogen-poor plants.

In most cases the nitrogen-fixing bacteria enter root cells of a target plant and multiply. Eventually a lump or a **nodule** forms at the site of bacterial entry. At this point the bacteria make use of a special set of enzymes known as the

nitrogenase complex. This complex involves six proteins and the activities of two distinct enzymes, one of which is called simply **nitrogenase** while the other is called **nitrogenase reductase.** These proteins and enzymes work together to satisfy the nitrogen requirements of the growing plant by carrying out the complex process of nitrogen fixation within the root nodules of that plant.

The bacterial species able to fix nitrogen include ***Rhizobium,*** which is often associated with the roots of legume plants such as beans and peas; ***Azospirillum,*** associated with the roots of grasses such as rye and wheat; and ***Frankia,*** often associated with the roots of some shrubs and trees. It is interesting to note that despite the absolute requirement of nitrogen fixation for successful growth, plants depend on associated bacteria to carry out this task.

Although other bacteria also fix nitrogen, *Rhizobium* shares that fixed nitrogen with its associated plant most efficiently. For example, although *Azospirillum* fixes nitrogen on the roots of grasses, it does not efficiently share that modified nitrogen with the associated plants. Therefore, despite the presence of nitrogen-fixing bacteria, the growth of crops such as corn, wheat, and rice quickly depletes the useful supply of nitrogenous compounds available in the soil, thus limiting the growth of the plants themselves. In order to continue to grow these crops, an exogenous source of nitrogen compounds must be added to the soil, usually by adding generous amounts of a nitrogen-containing fertilizer. Although this practice allows continued crop growth, it also creates problems: the production of nitrogen fertilizer adds extra expense to crop growth—an expense which cannot always be met, especially in poor and developing countries. In addition, the application of chemical-based fertilizers to growth areas can contaminate water supplies as some of that fertilizer washes away by the simple act of crop irrigation. This situation can adversely affect supplies of drinking water, and it can decrease aquatic animal populations as the growth of algae and other plants is enhanced by the presence of excess fertilizer-derived nitrogen in lakes and rivers.

For these reasons genetic engineers would like to find ways to increase supplies of fixed nitrogen, making them available to a wide variety of plants. This type of research takes many routes, some more promising than others.

In one avenue of examination, scientists would like to modify the factors which cause host-restriction in the relationship between *Rhizobium* and legume plants. This may take place either by altering the genes of the *Rhizobium* itself or by altering the genes of the associated plant. Although research continues in these areas, it is hampered by the extreme complexity of the genetic systems involved in the symbiotic relationship between the plant and the bacteria which provide fixed nitrogen.

In a second avenue of examination, researchers are attempting to determine why some bacteria, while capable of fixing nitrogen, do not share this valuable product with their associated plants. If this characteristic results from a specific gene function, genetic engineers should be able to identify, locate, and modify this gene or gene family and its controlling mechanisms.

Although these and other research questions are being examined in the area of nitrogen fixation, answers and solutions to problems depend on the ability of researchers to unravel the extremely complex processes of nitrogen fixation and on the ability of any genetic modifications to survive and produce in the field.

RESISTANCE TO FROST FORMATION

If the temperature falls below 32°F (0°C), ice crystals form and water freezes. Ice crystals, forming in the interior of a cell, often destroy the cell wall with the result that cell death soon follows. This phenomenon can be graphically illustrated by freezing a piece of firm crisp lettuce. As the temperature of the lettuce drops, ice crystals form and disrupt the delicate cells of the leaf. When the lettuce is later allowed to thaw, only a soggy, limp leaf remains where once a crisp leaf existed. A similar effect can be seen on growing plants which are subjected to freezing temperatures.

The presence of the common bacterium **Pseudomonas syringae** facilitates ice crystal or frost formation on a plant. *P. syringae,* found on many types of plants, contains a protein which acts as a site of nucleation for the formation of ice crystals at temperatures of approximately 32°F (0°C). In contrast, when *P. syringae* or other nucleating agents are absent, ice crystals do not form until the temperature drops to approximately 20°F (−7°C).

Scientists found that the nucleation ability of *P. syringae* results from a single gene in the bacterial genome. By removing this gene, scientists created a strain of bacteria which, when applied to growing plants, does not provide the necessary site of nucleation for frost formation even when the temperature drops briefly to as low as 23°F (−5°C). This genetically engineered bacterial strain, called **ice minus** and developed at the University of California at Berkeley, may someday be widely used to decrease or prevent frost damage to a variety of crops. Widespread usage of the engineered ice-minus bacterium awaits further efficacy studies as well as a thorough examination of the effects of releasing such an engineered organism into the general environment.

PLANTS WITH MORE COMPLETE PROTEINS

Protein molecules are composed of varying arrays of twenty different amino acids, of which the human body can synthesize twelve. The remaining eight amino acids, called **essential amino acids,** must be provided to the body by ingestion. This means that people must eat foods containing these eight amino acids to provide the complete proteins necessary for growth. All of the eight essential amino acids are present in a wide variety of animal products, including red meat, poultry, and milk. In contrast, however, no source of plant food contains adequate supplies of all eight of the amino acids—they all lack at least one. Consider, for example, the lowly bean. While beans have more than enough of the essential amino acid **lysine,** they lack the amino acid **methionine.** Wheat and rice, on the other hand, both contain suboptimal levels of lysine while containing a sufficient amount of methionine. When we consider that the majority of the world's human population exists on a diet which rarely, if ever, contains meat, we realize how important it is for people to be able to get sufficient quantities of each amino acid from a plant-based diet. While countless vegetarians have done so, the task would be significantly easier if plants contained complete proteins.

Scientists are currently using the knowledge gained through biotechnology in an attempt to alter the genes of a variety of plant proteins. For example, perhaps the genes of the bean could be altered so as to encode a protein which contains sufficient quantities of methionine: if the gene for **phaseolin**—the primary protein molecule of a bean—could be altered to contain codons which specify the amino acid methionine without altering the overall structure or growth pattern of the plant, phaseolin could become a complete protein.

The plant modifications described above are only examples of the many areas which researchers are examining. In addition, scientists hope to generate plants which have an increased tolerance for salinity or growth in salt water, a tolerance for highly acidic or highly basic soil conditions, resistance to viral infections, increased ability to store nutrients, and the ability to grow in the presence of various soil contaminants such as metals, to name but a few.

ANIMAL PRODUCTION

For many centuries farmers relied on two methods to increase their animal stock: either they purchased additional animals or they bred their existing animals. While ready cash limited the first option, the second option depended on the number of offspring which a single animal could bear. Biotechnology has changed the second limitation. Scientists can now use injections of exogenous hormone, either natural or recombinant, to induce **superovulation**—the production of a greater-than-normal number of eggs by a single female animal at a single time. While a normal cow might produce one, or at most two, eggs during a single ovulatory period, an animal subjected to superovulation might produce as many as eight or even ten eggs. These eggs can then be fertilized **in vitro**—in a laboratory dish—and implanted into host mothers. Thus, a single prize animal can give rise to eight or ten offspring instead of the more traditional single offspring. It is clear that the gene pool of animal populations can be greatly improved in this manner (Figure 12.3).

Genetically Engineered Vaccines Used With Livestock

One of the most important advances in agriculture has been the development of vaccines to treat diseases such as **hoof-and-mouth disease, scours, trypanosomiasis, rabies,** and such parasites as **tapeworms** and **liver flukes.** At one time an outbreak of some of these diseases would have forced farmers to kill entire herds; now animals can be vaccinated against them. The recombinant subunit vaccines (Chapter Eleven) currently in use, thanks to the techniques of genetic engineering, cannot cause an outbreak of the very disease which they seek to prevent, are relatively inexpensive, and do not require yearly booster shots.

Figure 12.3
Many of the cattle we see today result from one or more different applications of genetic engineering.

Genetically Altered Fish May Soon Feed Many Hungry People

In 1985 a group of scientists at the Academy of Sciences in Wuhan, China, first inserted copies of the gene which encodes human growth hormone into the eggs of goldfish. The engineered eggs then developed into mature fish. Analyses showed that approximately fifty percent of those fish incorporated the human growth hormone gene into the endogenous genetic material. Many of those transformed fish grew to sizes which were as much as four times the size of untransformed goldfish.

In the years since 1985 researchers in many countries have attempted similar experiments including the insertion of genes from a variety of species, ranging from human to bovine and murine, into a variety of fish, including salmon, rainbow trout, catfish, and carp. Scientists have attempted not only to increase the size and therefore, the food value of the fish, but also to transfer the gene that allows some fish to survive in frigid waters into fish which are more commonly found only in more temperate areas. Many scientists would also like to use the techniques of genetic engineering to create fish better able to survive in increasingly polluted waterways, disease-resistant fish, and fish that could withstand warmer as well as cooler than normal temperatures. It is hoped that this type of research will provide greater food resources for a hungry world.

USE OF ENGINEERED MOLECULES IN ANIMALS

In 1985 the FDA approved the experimental use of genetically engineered bovine growth hormone to increase milk production in cows. Previous research showed that use of the hormone increased milk production by as much as twenty-five percent as compared with cows not treated with the hormone. Engineered hormones now increase the ratio of muscle to fat in beef cattle and also induce an earlier production of functional sperm in bulls used as donors in the process of artificial insemination. Similarly, other engineered molecules affect change in a variety of animals: treated pigs produce meat that contains less fat, cattle are treated to prevent shipping fever, and treated chickens can grow more quickly and produce greater numbers of eggs.

REVIEW AND SUMMARY

1. Although biotechnologists working with plant cells benefit by the fact that such cells can be cloned easily and some plant species can be grown from single cells, genes have not always been cloned easily in plant host cells. The major problems with the use of plant cells as hosts in cloning protocols was the lack of a suitable cloning vector to carry inserted genes into host cells, the presence of a cell wall, the long generation time, and the relatively large genome of most plants.

2. The bacterium *A. tumefaciens* has the ability to cause the formation of plant tumors called **crown galls.** This bacterium contains a plasmid molecule called Ti. The **Ti plasmid** has been successfully engineered to allow it to play a role in the cloning of genes in plant host cells. In order to use the Ti plasmid in this way, the endogenous tumor-inducing genes must be inactivated, and the foreign gene which is to be cloned must be inserted into the T DNA region of the Ti plasmid molecule.

3. *A. tumefaciens* is restricted in its host range to dicotyledonous plants. In order to transform cells of monocotyledons, scientists have developed techniques that do not rely on the infective ability of this bacteria. One such technique, that of **microinjection,** uses a microneedle to inject naked DNA into a host cell. Another technique involves the brief application of high voltage electricity and is called **electroporation.** Electroporation causes the formation of transient pores in the cell membrane through which naked DNA molecules may migrate. A third technique, called **particle bombardment,** involves the production of DNA-coated micropellets which are shot through the cell wall.

4. A wide variety of genes, including those which confer resistance to the herbicide glyphosate, have been inserted into different types of plant cells. Other experiments have involved the engineering of bacteria which are then applied to a plant to provide assistance with nitrogen fixation, resistance to pests, or resistance to ice crystal formation. The protein quality of plants is also being modulated by genetic engineering.

5. The manipulations of biotechnologists also affect animals. Genetically engineered subunit vaccines have protected animals from a wide variety of diseases, including hoof-and-mouth disease and trypanosomiasis. Animals have had their growth and their fertility modulated with the injection of synthetic hormones.

For further reading

Baskin, Y. 1984. *The gene doctors: Medical genetics at the frontier.* William Morrow and Company, Inc. New York.

Cohen, S. N., Chang, A. C. Y., Boyer, H. W. 1973. Construction of biologically functional bacterial plasmids *in vitro. Proc. Nat. Acad. Sci. USA.* 70:3240–3244.

Darnell, J. E. 1985. RNA. *Scientific American.* 253:68–87.

Dickerson, R. E. 1983. The DNA helix and how it is read. *Scientific American.* 249:94–111.

Doolittle, R. F. 1985. Proteins. *Scientific American.* 253:88–99.

Drlica, K. 1984. *Understanding DNA and gene cloning. A guide for the curious.* John Wiley and Sons. New York.

Felsenfeld, G. 1985. DNA. *Scientific American.* 253:58–67.

Glover, D. M. 1986. *Gene cloning: The mechanisms of DNA manipulation.* Chapman and Hall. New York.

Hackett, P. B., Fuchs, J. A., Messing, J. W. 1988. *An introduction to recombinant DNA techniques. Basic experiments in gene manipulation.* 2d. ed. Benjamin/Cummings Publishing Co., Inc. Menlo Park, CA.

Hall, S. S. 1990. James Watson and the Search for Biology's 'Holy Grail.' *Smithsonian* (February 1990):40–49.

The Human Genome Map 1990. *Science* 250:262a–262p.

Koshland, D., ed. 1986. *Biotechnology: The renewable frontier.* AAAS. Washington, D.C.

Marx, J. L. ed. 1989. *A revolution in biotechnology.* Cambridge University Press. Cambridge.

Montgomery, G. 1990. The ultimate medicine. *Discover* (March 1990): 60–68.

National Research Council. 1988. *Mapping and sequencing the human genome.* National Academy Press. Washington, D.C.

Novick, R. P. 1980. Plasmids. *Scientific American.* 243:103–127.

Oliver, S., and Ward, J. 1985. *A dictionary of genetic engineering.* Cambridge University Press. New York.

Präve, R., Faust, U., Sittig, W., and Sukatsch, D. A., eds. 1989. *Basic biotechnology: A student's guide.* VCH Publishers. New York.

Ptashne, M. 1989. How Gene Activators Work. *Scientific American.* 260:41–47.

Rawn, D. J. 1989. *Biochemistry.* Neil Patterson Pub. Burlington, NC.

Sambrook, J., Fritsch, E. F., and Maniatis, T. 1989. *Molecular cloning: A laboratory manual.* Vol 1–3. Cold Spring Harbor Laboratory Press. New York.

The science of AIDS: Readings from Scientific American. 1989. W. H. Freeman and Company. New York.

Thornton, J. I. 1989. DNA profiling: New tool links evidence to suspects with high certainty. *Chemical and Engineering News.* 20 (November):18–30.

Watson, J. D. 1980. *The double helix: A personal account of the discovery of the structure of DNA.* W. W. Norton and Co. New York.

Watson, J. D., and Tooze, J. 1981. *The DNA story: A documentary history of gene cloning.* W. H. Freeman and Co. San Francisco.

Watson, J. D., Tooze, J., and Kurtz, D. T. 1983. *Recombinant DNA: A short course.* Scientific American Books. New York.

Weaver, R. F. 1984. Beyond Supermouse: changing life's genetic blueprint. *National Geographic.* 166:818–847.

Weinberg, R. A. 1985. The molecules of life. *Scientific American.* 253:48.

White, R., Lalouel, J. M. 1988. Chromosome mapping with DNA markers. *Scientific American.* 258:40–49.

Williams, J. G., and Patient, R. K. 1988. *In focus: Genetic engineering.* Ed. by D. Rickwood, IRL Press. Oxford and Washington, D.C.

Glossary

ALU Sequence nucleotide sequence which contains the *Alu*I recognition site and which is repeated between 100,000 and 300,000 times in the human genome and therefore acts as a marker for human DNA sequences.

Anticodon the 3-nucleotide region of a tRNA molecule, complementary to a specific mRNA codon, that therefore specifies the amino acid to be carried by that tRNA molecule.

Autoradiography detects the location of radioactive isotopes in a gel or other substrate. The labelled sample is placed against undeveloped film. Radioactive emissions then develop the film in the immediate area surrounding the isotope particle. Thus, the developed spot corresponds to the label location in the substrate being examined.

Bacteriophage a virus that infects bacterial cells. Also called a **phage.**

Base a compound which forms an integral part of a nucleic acid. There are five common bases: the **pyrimidines**—thymine (DNA only), cytosine, and uracil (RNA only), and the **purines**—guanine and adenine (Figure 1.9). A base chemically bound to a sugar is referred to as a **nucleoside,** while a nucleoside bound to a phosphate group is a **nucleotide.**

CAAT Box a nucleotide base sequence located approximately 75 base pairs upstream of a eukaryotic transcriptional start site and which appears to enhance binding of RNA polymerase II to the promoter.

Catabolite Activator Protein (Cap) a bacterial protein which binds to a site located in or near the promoter and enhances binding of RNA polymerase to that promoter. As a result, the expression of associated genes is stimulated.

cDNA DNA synthesized from an mRNA template using the enzyme **reverse transcriptase.** cDNA contains only exon sequences while eukaryotic genomic DNA contains both exon and intron sequences.

Central Dogma states that hereditary information travels in one direction only: from DNA to RNA to polypeptide product.

Chromosome in eukaryotes: rod-like structure composed of DNA and protein. In prokaryotes: circular structure composed solely of DNA. Chromosomes are divided into functional subunits called **genes.**

Clone (noun) a population of genetically identical molecules, cells, or organisms.

Clone (verb) the production of an identical population of cells or, usually using genetic engineering techniques, of a population of identical DNA fragments.

Codon see Genetic Code.

Competent Cells cells that are able to take up exogenous genetic material.

Consensus Sequence a highly conserved sequence of nucleotides which is found in most examples of any particular type of genetic element. For example, the sequence TATA is almost always found within eukaryotic promoters. Thus, the sequence TATA is the consensus sequence for a genetic element which has come to be known as the **TATA box.**

Cosmid a cloning vector which contains the lambda phage *cos* site incorporated into plasmid DNA sequences which often contain at least one selectable marker and an origin of replication. Cosmids accept the insertion of relatively large DNA fragments of up to approximately 40,000 base pairs.

Deoxyribonucleic Acid (DNA) a nucleic acid molecule composed of two complementary strands of nucleotides which are wound about one another in a double helix formation. DNA is the primary genetic material of all living organisms.

Downstream describes the relative positions of two loci on a single DNA fragment. When A is downstream of B, A is on the 3' side of B.

Drug Resistance the ability of a cell to resist the lethal effects of one or more specific drugs, usually antibiotics.

Enhancer Sequence nucleotide sequence, located as many as several thousand base pairs in either direction from the target gene, which enhances transcription of that gene.

Enzyme a protein which speeds up a chemical reaction without itself being consumed in the reaction.

Eukaryote an organism composed of one or more cells, each of which contains a membrane-bound nucleus.

Exon the coding sequences of a gene. The exon sequences are represented in the final gene product.

Expression Library a library constructed with expression vectors as the cloning vehicle. Such a library can be screened by the identification of specific gene products.

Expression Vector a vector containing all of the genetic regulatory elements necessary for expression of the inserted exogenous gene.

Gel Electrophoresis separation of a population of heterogeneous molecules by passage through a molecular sieve in the form of a gel. Molecules are pulled through the pores of the sieve by an electric current. In general, smaller molecules move the farthest into the gel while larger molecules remain closer to the origin.

Gene a unit of hereditary information located on a chromosome.

Genetic Code the language which allows conversion of genetic information contained in an mRNA molecule consisting of an arrangement of only four different nucleotides, into a polypeptide product. The words of the genetic code are called **codons.** Each codon consists of three adjacent nucleotides in an mRNA molecule. Of sixty-four possible codons, sixty-one specify a single amino acid, and three act as punctuation. Most of the twenty-one amino acids are specified by more than one codon. No single codon, however, specifies more than one amino acid.

Genetic Complement the set of chromosomes contained within any one particular cell.

Genetic Engineering the *in vitro* manipulation of DNA molecules to create combinations of DNA sequences, including genes, which are not normally found in nature.

Genetic Library a population of recombinant DNA molecules, each of which contains a piece of target DNA. All of the target DNA fragments in a library, when taken together, make up the total genome of a target organism.

Genomic DNA DNA collected directly from a cellular source prior to any genetic processing. Eukaryotic genomic DNA contains both intron and exon sequences.

Insertional Inactivation the inactivation of a gene due to the insertion of exogenous genetic material into that gene.

Intron (Intervening Sequence) the non-coding sequences located within eukaryotic genes. Although intron sequences are transcribed into RNA molecules, they must be removed prior to translation of that RNA in order to produce a functional, or potentially functional, polypeptide product.

Intron Splicing the excision of mRNA fragments, which correspond to non-coding DNA introns, and the ligation of the remaining mRNA fragments to form a single molecule.

Isoschizomer a pair of restriction enzymes specific for the same recognition sequence.

Lambda Phage a bacteriophage which is the basis for many successful gene cloning vectors. Lambda genetic material consists of a double-stranded DNA molecule with 5′ twelve-base-pair sticky ends, known as *cos* sites, which permit circularization of the DNA molecule. Lambda phage can undergo both lytic and lysogenic infective cycles.

Linker a sequence of deoxyribonucleotides constructed to contain a known restriction enzyme recognition site. Linkers can be ligated to the end of DNA fragments to add desired restriction sites which can then be used in cloning protocols.

Lysogenic Infection an infective process characterized by the incorporation of the DNA of the infecting phage into the host cell chromosome. Once incorporated, the phage DNA replicates along with the host DNA. The incorporated phage DNA is relatively inactive, thus permitting the host cell to continue fairly normal life processes.

Lytic Infection an infective process characterized by the uncontrolled replication of bacteriophage DNA which has entered a host cell. The newly synthesized phage DNA molecules are packaged into phage heads which are, in turn, incorporated into phage particles. Production of a large number of new phage particles results in lysis of the host cell and release of mature phage particles which can go on to infect neighboring bacterial cells.

Microinjection a technique in which exogenous DNA is injected through a microscopic needle into a recipient cell.

Minimal Medium bacterial growth medium consisting of defined chemical components.

Mutation an unusual change in DNA base sequence leading to a change in hereditary information.

Nucleoid Region area in prokaryotic cell which contains the genetic material.

Nucleoside nucleic acid subunit consisting of a sugar and a nitrogenous base. The sugar in RNA nucleosides is ribose while the sugar in DNA nucleosides is deoxyribose.

Nucleotide nucleic acid subunit consisting of a sugar, a phosphate group, and a nitrogenous base. The sugar in RNA nucleotides is ribose while the sugar in DNA nucleotides is deoxyribose.

Nucleus the membrane bound-region in eukaryotic cells which encloses the genetic material.

Operator a nucleotide sequence located very close to, or overlapping, the promoter, and to which a repressor molecule may bind and cause suppression of gene expression.

Operon a group of genes with related functions whose expression is coordinated with one another. Commonly found in bacteria and phage particles.

Plasmid generally a circular, double-stranded DNA molecule which is separate from the chromosomal material and which is able to replicate independently of the chromosomal material. Plasmids are most often found in bacterial cells. Many cloning vectors are derived from the genetic manipulation of plasmids.

Poly-A Tail sequence of adenine bases, ranging from approximately 50 to 250 residues in length, which is attached to the 3' end of most eukaryotic mRNA molecules during post-transcriptional processing.

Polymerase an enzyme whose function is to assist in the joining or polymerizing of a series of subunits to one another.

Polypeptide a chain of amino acids joined by peptide bonds. Some polypeptides are functional proteins while others are protein subunits.

Pribnow Box a sequence of nucleotides common to prokaryotic promoters and located approximately ten base pairs upstream of the transcriptional start site.

Probe a molecular tool used to screen a gene library or other mixed population of molecules. The target is fished or selected from the heterogenous population by its complementarity to the probe.

Prokaryotic Cell a cell which lacks a membrane-bound nucleus, e.g. bacteria. Prokaryotic genetic material is in the form of a single, circular, DNA molecule and is arranged in **operons.**

Promoter a specific sequence of nucleotides which acts as the binding site for RNA polymerase and therefore indicates the correct transcriptional start site.

Protease a non-specific name for an enzyme which functions to degrade protein molecules.

Recognition Sequence the particular sequence of base pairs to which a restriction enzyme binds. Type II restriction endonucleases go on to cleave the DNA molecule at a site located within the recognition sequence.

Recombinant DNA DNA molecules consisting of two or more fragments which would not normally be found next to one another in nature. The fragments which make up a recombinant DNA molecule are often from different organisms and can only be joined by the techniques of genetic engineering.

Replication the synthesis of new DNA strands using pre-existing DNA strands as templates.

Repressor Protein a protein which binds to a prokaryotic operator and thereby causes suppression of the expression of associated structural genes.

Restriction Enzyme an enzyme which recognizes and binds to a particular sequence of bases within a DNA molecule. After binding, the enzyme cleaves the backbones of the DNA molecule.

Restriction Map a schematic representation of the location of various restriction enzyme recognition sequences on a DNA molecule.

Restriction Fragment Length Polymorphism (RFLP) homologous DNA fragments of different lengths which occur due to variations in the DNA sequences on homologous chromosomes. RFLPs are useful as genetic markers and can be used to trace the passage of certain traits through multiple generations of the same family.

Retrovirus a class of viruses that contains RNA as its genetic material. Retroviruses function by attaching to a host cell and injecting genetic materials and enzymes into that cell, causing the host to synthesize a DNA copy of the viral RNA molecule. Retroviruses are important in some types of cancers and in AIDS.

Reverse Transcriptase an enzyme found in retroviruses which directs synthesis of a DNA copy of an RNA template.

Ribonucleic Acid (RNA) nucleic acid composed of ribonucleotides which, in turn, are composed of a ribose sugar plus a phosphate group and a nitrogenous base. There are three major classes of RNA: mRNA (messenger RNA), which carries hereditary information from a DNA template; rRNA (ribosome RNA), a component of ribosomes; and tRNA (transfer RNA), which carries amino acids to growing polypeptides during translation.

RNase also known as **ribonuclease,** this class of enzymes functions to degrade RNA molecules.

Screen analysis of a population to determine which members fit a specified criterion. Used to identify particular inserted genes in a transformed population. A genomic library is **screened** to identify a particular target genetic insert.

Selection growth of cell populations under conditions chosen to allow growth of one segment of the population while preventing growth of another segment. Often used to identify a subpopulation of transformed cells from within a larger parent population. A population of cells which has been treated with calcium precipitates of DNA undergoes **selection** to determine which cells have actually taken up and retained exogenous DNA.

Shine-Dalgarno Sequence (Ribosome Binding Site) a particular nucleotide sequence located within an mRNA molecule which is complementary to a specific ribosomal site and to which a ribosome will bind. Important in establishing the correct reading frame for translation.

Shotgun Cloning a process in which random fragments of a target genome are inserted into cloning vectors which can then be converted into a gene library. The library can subsequently be screened for particular target genes.

Shuttle Vector a cloning vector able to replicate in two different organisms and which can therefore shuttle DNA inserts between organisms.

Simian Virus 40 (SV40) a virus which normally infects monkey cells and which is the basis for a series of useful cloning vectors. SV40 carries out both lytic and lysogenic infections.

S1 Nuclease a nuclease which preferentially degrades RNA or single-stranded DNA.

Sticky Ends single-stranded ends on DNA fragments, often left by the action of restriction endonucleases, which are complementary to one another and can therefore be used to join different fragments in the formation of a recombinant molecule.

Synapsis the pairing of homologous chromosomes during the cell-division process of meiosis.

TATA Box a eukaryotic-promoter-associated nucleotide sequence located approximately 25 base pairs upstream of the transcriptional start site and which facilitates, but is not necessary to, the process of transcription. The TATA box is analogous to the prokaryotic Pribnow box.

Transcription the synthesis of RNA molecule complementary to a DNA template. Transcription results in the transfer of hereditary information from a DNA molecule to an RNA molecule.

Transduction the transfer of DNA from one bacterium to another via the processes of release of infecting recombinant phage from one cell and the subsequent infection of other cells.

Transfection the transformation of bacterial host cells with a phage recombinant cloning vector.

Transformation the process in which exogenous genetic material is taken up by an appropriate host cell. Cells which have undergone transformation are said to be transformed.

Translation the synthesis of polypeptide molecules, directed by the hereditary information contained within an mRNA molecule.

Vector sometimes called a **cloning vector**. A DNA molecule capable of acting as a vehicle in molecular cloning. The fragment to be cloned is inserted into the vector which is in turn inserted into an appropriate host organism. The most useful vectors contain one or more selectable markers, one or more unique restriction sites, and, in addition, are capable of autonomous replication.

Watson-Crick Base Pairing the pattern of complementary base pairing seen in DNA molecules in which guanine pairs with cytosine while thymine pairs with adenine.

Credits

I.1: Bettmann Archives; I.2 & 3: National Gallery of Medicine; I.4a: Bettmann Archives; I.4b: Cold Spring Harbor Laboratory

Chapter 1

1.1a: © Mike and Elvan Habicht/Animals Animals/Earth Scenes; 1.1b: © Cabisco/Visuals Unlimited; 1.1c: © Pat Crowe/Animals Animals; 1.1d: © Leonard Lee Rue III/Animals Animals/Earth Scenes; 1.3a: © John Cunningham/Visuals Unlimited; 1.4a,b: Kleeberg Cytogenetics Laboratory

Chapter 2

2.1: © Don W. Fawcett/Photo Researchers, Inc.; 2.2: Courtesy of Sandra MacKenzie/Marilyn Holmes Photography; 2.3a: © Dr. R. G. Kessel; 2.3b: © S. L. Flegler/Visuals Unlimited

Chapter 3

3.1a: © Visuals Unlimited; 3.1b: © U.S. Air Force Photo/Photo Researchers, Inc.; 3.2: © Bob Coyle; 3.9: © Mike Siluk/The Image Works

Chapter 4

4.1a: © SIU/Photo Researchers, Inc.; 4.2a: © Bob Daemmrich/The Image Works; 4.2b: © Paula Wright/Animals Animals; 4.3: © Richard Rodewald/Biological Photo Service; 4.4: © Henry Aldrich/Visuals Unlimited; 4.5a: © Dr. Richard P. Novack *Scientific American,* Dec. 1980, p. 104; 4.5b: From Ruth Kavenoff, Lynn C. Klotz, and Bruno H. Zimm, Symopsia on Quantitative Biology (Cold Spring Harbor) 38 (1973):4; 4.5c: © Paul Heidger; 4.5d: Courtesy James Paulson *Cell* 12:823, 1977; 4.5e: © Dr. Keith Porter; 4.5f: © Biophoto Associates/Science Source/Photo Researchers, Inc.; 4.6: © Jack M. Bostrack/Visuals Unlimited

Chapter 5

5.2a: © Richard Anderson; 5.2b: © Raymond B. Otero/Visuals Unlimited; 5.4: © Fred Marsik/Visuals Unlimited; 5.5: © Science VU/Visuals Unlimited

Chapter 6

6.1: © Robert Perron/Photo Researchers, Inc.; 6.9a: © Lee D. Simon/Photo Researchers, Inc.

Chapter 7

7.1: © Dion Ogust/The Image Works; 7.2: © Dr. Tony Brain/Science Photo Library/Photo Researchers, Inc.; 7.3: © T. S. Anderson; 7.4a: © Bruce Iverson/Visuals Unlimited; 7.4b: © E. C. S. Chan/Visuals Unlimited

Chapter 8

8.1: © Paula Wright/Animals Animals; 8.5: © Dr. Jan-Erik Edstrom, *Developmental Biology* vol. 91:131–137, fig. 1b, 1982; 8.6: © Behnke/Animals Animals

Chapter 9

9.1: © Mark Antman/The Image Works; 9.2a: © David Scharf/Peter Arnold, Inc.

Chapter 10

10.1: © Visuals Unlimited; 10.2: © M. M. Perry and A. B. Gilbert, *Journal of Cell Science* vol. 39:257–272, 1979; 10.3: © Hank Morgan/Photo Researchers, Inc.

Chapter 11

11.1: © Leonard Lessin/Peter Arnold, Inc.; 11.4: © Baylor/Medicine/Peter Arnold, Inc.; 11.5: Courtesy of Genetech, Inc.; 11.6: The American Lung Association; 11.7: © Robert Gallo/National Institutes of Health

Chapter 12

12.1: © Richard Choy/Peter Arnold, Inc.; 12.3: © George F. Godfrey/Animals Animals

Index

A

Acetabularia, phenotypes of, 4–7
Acquired Immune Deficiency Syndrome (AIDS), 1, 11, 214–17, 219
Activator proteins, 146–47, 150–51, 154
Adenine, 13, 14, 16–17, 29, 35
Adenosine deaminase deficiency (ADD), 204–5
Affinity chromatography, 98–99
Agarose gel electrophoresis, 75–81, 131, 135
Agrobacterium tumefaciens, 222, 230
AIDS, 1, 11, 214–17, 219
Albino, 130
Alcohol dehydrogenase, 168
Aliquots, 85–87
Alkaline phosphatase, recombinant vectors and, 127–28, 138
Alu sequence, 151, 179–80, 190
Alzheimer's disease, 200, 209
Amino acids
　of DNA polynucleotides, 13–14, 16–17, 23, 25
　of plants, 227–28
　in protein synthesis, 37–39
　of RNA, 29, 34–36
Amino glycoside 3′ phosphotransferase, 183
Ampicillin, 106–7
Amplification, 107–8
Anderson, W. French, 204, 205, 209
Aneuploid mutation, 44
Animals, genetic engineering of, 228–30
Antibodies, 136
　anti-DNA, 197
　anti-HLA, 186
Antibody probes, 130, 136–37
Anticodon, 36, 37
Anti-DNA antibody, 197
Antigens, 132
　human leukocyte, 185–86
　of vaccines, 211–12
Anti-HLA antibody, 186
Antisense (non-coding) strand, 30, 31, 33, 34
alpha-1 antitrypsin (AAT), 210–11, 217–18
Arber, Werner, xiii, 62
Aspartame, 192
Autonomously replicating sequence (ARS), 164, 168–69
Autoradiography, 87, 90, 135, 138
Azospirillum, 226

B

Bacillus, competence of, 119
Bacillus thuringiensis, 224–25
Bacteria, 54, 56, 59, 140. *See also* Prokaryotic cells; *names of specific bacteria*
　culturing of, 120–23
　nitrogen fixing, 225–27
Bacteriophages, 66, 123
　as cloning vectors, 108–13, 114
　lambda, 108–11, 112
　T2, 10, 12

Baculovirus, 225
*Bam*HI, 83
    in DNA cleavage, 65, 68, 69, 74–75
    in gene cloning, 97, 103, 105, 165–66
Bases of DNA, 13–15. See also Recognition sequences
    blunt and staggered, 67–70
    cleaving of, 63–65
    determining sequencing of, 85–90
    pairing of, 15–17
Beadle, George, xiii, 26, 27
Binary fission, 120
Biotechnology. See also Gene cloning; Recombinant DNA molecule
    in animal production, 228–30
    definition of, xi–xii
    in gene replacement therapy, 202–5
    genetically engineered hormones of, 205–7
    lymphokines in, 207–9
    mapping of genes in, 81–82, 196–202
    of plant cells, 221–28
    therapeutic enzymes of, 209–11
    vaccines of, 211–14
Blaese, R. Michael, 204, 205
Boyer, Herbert, xiii, 124
Brinster, Ralph, 177
Bromodeoxyuridine (BrdU), 178

C

CAAT box, 155, 159, 184
Cachexia, 209
Calcium chloride, 123
Calcium-mediated DNA uptake, 174, 190
Catabolite activator (CAP) proteins, 146–48, 158
Catabolite repression, 147–48
CD4, 215
cDNA. See Complementary DNA (cDNA)
cDNA probes, 132–33
Cells. See also Eukaryotic cells; Prokaryotic cells; names of specific cells
    competent, 119, 120, 123, 127
    hereditary material of, 10–12, 55–59
    phenotype of acetabularian, 4–7
    somatic and germ, 8–9, 51, 52
    transformed, 96, 123–26, 138
cen fragment, 164–65

Central Dogma, xiii, 28, 29
    in transcription, 30–36
    in translation, 36–39
Centrifugation, 61, 62
Chase, Martha, xiii, 10, 12
Chloramphenicol, 107–8
5-chloro-4-bromo-3-indolyl-beta-d-galactoside, 128
Chromatin, 56, 59
    fiber, 152
Chromatography, 98–99
Chromosomal puffs, 152–53, 154
Chromosomes. See also Genes
    diploid and haploid, 8–9
    genetic complement of, 43
    homologous, 45, 47
    mapping with, 197–98, 199, 218
    mutations of, 44–49
    polytene, 152–53
    of prokaryotic and eukaryotic cells, 58–60, 70
    of yeast cells, 162
Chromosome walking, 199, 218
Cleavage of DNA
    blunt and staggered bases in, 67–68
    restriction nucleases in, 62–67
Clone, 92
Clone panels, 196–97, 218
Cloning. See Gene cloning; Recombinant DNA molecule
Cloning vectors, 94
    characteristics of, 102–5, 114
    cosmid, 111–13
    identifying recombinant, 126–28
    phages as, 108–11
    plasmids as, 105–8, 111, 164–65, 168–69
        shuttle, 166–67
        Ti, 222
    viral, 180–83
Coding strand, 30
    polymorphic, 82
Codons
    effects of genetic mutations on, 49–50
    to specify amino acids, 35–36, 167–68
    start, 37
    in translation, 37–39
Cohen, Stanley, xiii, 124

Color-blindness, 84, 195, 196
Competent cells, 119, 120, 123, 137
Complementary base pairing, 18, 19, 21, 30, 31
Complementary DNA (cDNA), 11
    as donor genome, 113–14
    libraries of, 117–18, 138
    probes, 132–33
    synthesis of, 100–102
Complement mutations, 166, 169
Concatemer, 110
Conneally, Michael, 84
Consensus sequence, 146
Corepressors, 145, 158
Cosmid vectors, 111–13, 14
cos sites, 108, 109, 110, 111, 112
Cowpox, 211, 212, 213–14
Crick, Francis, xiii, xv, 15
Cri-du-chat syndrome, 47
Crown gall, 222, 230
Crystal, Ronald, 211
Cultures
    mammalian tissue medium, 173
    phases of growing bacterial, 121–23
    screening of, 130–37
    selective media, 124, 125, 126, 128, 173, 179, 180
Cyclic adenosine monophosphate (cAMP), 147, 158
Cystic fibrosis, 200, 209, 211
Cytokinesis, 8
Cytosine, 13, 14, 16–17, 29, 35

D

Dalgarno, Larry, 148
Degradation in gene regulation, 156
Denaturing of DNA, 60, 85–86, 87, 90, 133
Deoxyribonucleic acid. *See* DNA
Deoxyribose, composition of, 13–15
Diabetes, 206
Dicotyledons, 222–23, 231
Dideoxynucleotides (ddNTPs), 88–90
Diploid cell, 8
Disaccharides, 141, 147
DNA, xiv. *See also* Recombinant DNA technology
    cell phenotype and, 4–7

    of chromosomes, 7–9, 56
    cleavage of, 62–67
    cloning of, 92–115
        gene libraries for, 116–38
        in mammalian hosts, 170–91
        in yeast cells, 160–69
    in eukaryotic and prokaryotic cells, 58–59
    formation of recombinant, 67–70
    gel electrophoresis of, 75–81
    genomic and plasmid isolation of, 60–61
    mapping, 81–82, 196–202
    mutations in replication of, 42–56
    polymorphisms, 82–85
    replication of, 18–21
    sequencing, 85–90
    as storehouse of heredity, 10–12
    structure of, 13–17, 28–29
    transcription into RNA, 30–32
DNA ligase, xiii, 69–70
DNA polymerase, 19, 43, 101
Donor genome
    cDNA and genomic DNA as, 100–102, 103, 113–14
    cloning vectors for, 102–13
    detecting transformed cells with, 124–26
    growing host cells for, 119–23
    selection and preparation of, 96–99
Down's syndrome, 44, 45
Drosophila, 152–53, 185
Drug resistance, 59
Dwarfism, 205

E

Early genes, 181
Ecdysone, 153
*Eco*RI
    in DNA cleavage, 65, 67, 68, 69
    in gene cloning, 97, 103, 112, 165–66
    of pSC101, 124–26
Electrophoresis. *See* Gel electrophoresis
Electroporation, 223, 230
Elongation, 40
    in transcription, 30, 31, 32
    in translation, 37, 38, 39
Emphysema, 210–11, 217–18
Endocytosis, 174, 175

Endonucleases
    in cleavage of DNA, 63–66, 74–75
    in cloning genes, 97, 102, 103–4, 105, 112, 113, 165–66
    in DNA mapping, 81–82
    of pSC101, 124–25
    type I and type II, 66–70, 71
Endotoxins, 172
Enhancer sequences, 155, 159, 184, 191
Enzymes. *See also* Restriction nucleases; *names of specific enzymes*
    in DNA replication, 19, 60, 69–70
    genes in production of, 26–27
    genetic engineering of therapeutic, 209–11, 219
    metabolic disorders and, 192–93, 195–96
Escherichia coli, xiii, 30
    detecting transformed cells of, 123–26
    in DNA cleavage, 62–63
    DNA replication in, 18, 20
    in gene expression, 141, 142–43, 147–48, 166–67, 172
    growing of cultures and strains of, 119–23
    for growth of therapeutic molecules, 206–7
    plasmid of, 105
    yeast analog of, 166–67
Ethidium bromide, 80
Eugenics, 203
Eukaryotic cells, 30
    cloning of
        in mammalian cells, 170–91
        in prokaryotic cells, 156–57
        in yeast cells, 160–69
    control of gene expression in, 145, 149–56, 184–85
    isolation of DNA from, 60–61, 101–2
    RNA of, 33, 34
    structure of, 56, 58, 71
Exogenous DNA, 123. *See also* Gene cloning
Exons, 33, 34, 101, 102
Exonucleases, 63
Expressed genes, 139. *See also* Gene expression
Expression gene libraries, 118–19, 135
Expression vectors, 118–19, 135, 157, 168

F

Fish, genetic engineering of, 229–30
Fluorescence-activated cell sorter (FACS), 185–89, 191, 197
Frameshift mutation, 49–50
Frankia, 226
Franklin, Rosalind, xiii, xv
Fusion protein, 157
Fusogen, 163, 173, 186

G

Galactose, 128, 141, 142
beta-galactosidase, 128, 141, 142, 144, 157
Gallo, Robert, 1
Gel electrophoresis, 90
    agarose, 75–81, 131, 135
    polyacrylamide, 85–87
Gene(s), 26. *See also* Chromosomes
    mutations of, 23–25, 44, 49–51, 166, 194–96, 200
    relationship of phenotypes to, 26–27
    transcription of, 30–32
    translation of, 36–39
Gene cloning, 92–95
    detection of transformed cells in, 123–26, 178–80
    libraries for, 117–18, 129–38
    in mammalian hosts, 170–91
    in plant cells, 221–28
    preparation of donor genome in, 102–13
    transformation of host cells in, 119–23
    vectors in, 96–102, 126–28, 164–65, 180–83
    in yeast cells, 160–69
Gene expression, 139–41
    control of eukaryotic, 149–56
    control of prokaryotic, 141–46
    of eukaryotic DNA in prokaryotic cells, 156–57
    interaction of promoters and CAP proteins in, 146–48
    in mammalian hosts, 184–91
    translational regulation in, 148–49
    in yeast hosts, 167–68

Gene libraries
    analyzing, 129–32
    cloning vectors in, 102–13
    constructing, 94–96
    maintaining, 137
    probes for screening, 132–37
Gene replacement therapy, 202–5, 218
Gene therapy, 208–9, 218
Genetic code, 34–36, 167–68
Genetic polymorphism, 82–83, 90–91
    restriction fragment length, 84–85, 199
Genomic DNA, 60, 99
    cloning in mammalian hosts, 170–91
    cloning in yeast cells, 162–69
    as donor genome, 100, 101, 102, 103, 113–14
    libraries of, 117–18
    mapping of human, 196–202
    Southern blotting analysis of, 133–35
Germ cells, 8, 51, 52, 203
G418, 183
Gilbert, Walter, 85
Glucocorticoid hormones (in gene expression), 184–85
Glucose, 141, 142, 147
Glycogen, 206
Glyphosate, 224, 231
gp120, 215, 219
Graham, F., 174
Guanine, 13, 14, 16–17, 29, 35
Guanosine monophosphate (GMP), 182
Gusella, James, 84, 200

H

*Hae*III, 67
Haemophilus aegyptius, 67
Haemophilus haemolyticus, 68
Haemophilus influenzae, 63, 65
Hairpin loop, 100, 101
Hairy cell leukemia, 208
Hammerling, Joachim, 4–5, 6
Haploid cells, 8
HAT medium, 179, 190
*Hba*I, 68
Heat inactivation, 60
Heat shock, 123, 185

Helper T cells, 215
Helper viruses, 181, 191
Hemoglobin, 23–25, 93, 99
Hemophilia, 195, 196
Herbicides, 224, 231
Heritable mutations, 51, 194–95. *See also* Inheritance
Hershey, Alfred, xiii, 10, 12
Hexosaminidase A, 195
HGPRT, 181–82
*Hin*dII, 63, 65, 81, 91
Histone proteins, 56, 59, 70, 152
HLA, 185–86
Homologous chromosomes, 45
Hoof-and-mouth disease, 213, 228, 231
Hormones
    gene expression with glucocorticoid, 184–85
    genetically engineered, 205–7, 219
Host cells, 96
    growing of *E. coli* as, 119–23
    lyophilization of, 137
    transformation of, 119–23, 124
Hot probe, 133
Housekeeping genes, 140
Human Genome Mapping Project, 200–202, 218
Human growth hormone (hGH), 205–6
Human health care. *See also* Mammalian hosts
    gene replacement therapy in, 302–5
    genetically engineered hormones in, 205–7
    lymphokines in, 207–9
    therapeutic enzymes in, 209–11
    vaccines in, 209–11
Human Immunodeficiency Virus (HIV), 11, 214–17, 219
Human leukocyte antigen (HLA), 185–86
Humulin, 207
Huntington's disease, 72, 84–85, 200
Hybridization
    in screening genes, 131, 134, 135
    of somatic cells, 172–74
Hybrid vectors, 111–13
Hydrolyzation, 121
Hydroxyl (-OH) group, 127–28
Hypoxanthine guanine phosphoribosyl transferase (HGPRT), 181–82

I

Ice minus, 227
Immune system, 136
  lymphokines of, 207-9, 219
Inducer, 144, 158
Inducible operons, 144, 158
  lac, 143-45, 148
Inducible promoters, 185
Ingram, Vernon, 27
Inheritance, 4, 5. *See also* Genes; Mutations
  DNA and, 10-12, 55-59
Initiation in translation, 37-39, 40
Insecticides, 224-25
Insertional inactivation, 106-7, 128, 138
Insertion mutations, 49, 52
Insulin, 54, 74, 99
  genetically engineered human, 205, 206-7, 219
Interferons, 168, 207-8
Interleukins, 208
Introns, 33, 34, 101, 102
  yeast, 162, 167
Islet cells, 96, 99
Isolation of DNA, chromosomal and plasmid, 60-61
Isoschizomers, 64

J

Jacob, Francois, 142
Jenner, Edward, 212

K

Kanamycin, 124, 125
Kaposi's sarcoma, 1, 216

L

lac operon, 143-45, 148, 157, 184
Lactase, 36-37
Lactose, 141, 142, 143-45
Lambda phage, 108-11, 112
Late genes, 181
Ligation of DNA, 33
  alkaline phosphatase inhibiting, 127-28
  with DNA ligase, 69-70

Lin, Stewart, xiii, 62
Linkers, 68
Lipids, 60
Liver flukes, 228
Lymphadenopathy, 215-16
Lymphocytes, gene replacement therapy with, 204
Lymphokine activated killer (LAK) cells, 208
Lymphokines, 207-9, 219
Lyophilization, 137
Lysate, 60
  cleared, 61
Lysine, 227
Lysogenic cycle of phages, 109, 110, 114
Lytic cycle of phages, 109, 110, 114

M

Macrophages, 215
Magic bullets, 209, 218
Malaria, 51
Mammalian hosts. *See also* Human health care
  calcium-mediated DNA uptake in, 174
  control of gene expression in, 184-85
  detection of cloned gene on surface of, 185-89
  microinjection of exogenous DNA into, 174-78
  selecting transformed, 178-80
  somatic cell hybridization in, 172-74
  use of, 170-72, 189
  viral cloning vectors and, 180-83
Mapping of genes
  chromosome walking, 199
  human genome, 200-202
  with markers and probes, 198-99
  restriction, 81-82
  RFLP, 199-200
  with somatic cell hybrids, 196-97
  by sorting chromosomes, 197-98
Maxam and Gilbert DNA sequencing, 85-87, 91
Media. *See* Cultures; Selective media
Meiosis, 45, 46
Melanoma, gene therapy for, 209
Memory B cells, 212
Mendel, Gregor, xii, xiii, 4, 5, 21

Mendelian Inheritance in Man (McKusick), 201
Meselson, Matthew, 18, 20
Messenger RNA (mRNA), 28, 29
    chromatographic isolation of, 97–99, 114
    probes in protein synthesis, 132–33, 138
    ribosome binding on, 148–49
    in synthesis of cDNA, 100–102
    in transcription, 33, 34, 155
    in translation, 37–39, 40
Metabolic genetic disorders, 192–93, 195–96, 218
Metallothionein gene, 177–78, 185
Methionine, 227–28
Methylene blue, 80–81
Methyl groups, 63, 66
Mice
    for manufacturing drugs, 170–71
    microinjection of genes in eggs of, 176–78
    beta$_2$-microglobulin, 185
Microinjection of DNA
    into fertilized mouse eggs, 176–78, 190, 203
    into plant cells, 223
    procedure for, 174–76, 191
    2 micron circle, 162–63
Miescher, Friedrich, xiii, xiv
Missense mutation, 49, 50, 52
Mitosis, 8
Monocotyledons, 223, 231
Monod, Jacques, 142
Monosaccharides, 141
Montagnier, Luc, 1
Morgan, Thomas Hunt, xiii, xiv
Mouse mammary tumor virus (MMTV), 184–85
Mutagens, 51–52
Mutations, 21, 43, 52–53. *See also* Gene cloning
    causes of, 51–52
    chromosomal, 44–49
    complement, 166
    gene, 49–51
    heritable, 51, 194–95, 200
        sickle cell, 23–25
    metabolic, 195–96
Mycophenolic acid, 182, 183

N

Nathans, Daniel, 81
Neomycin, 183
Neurospora, 26–27
Neutrophil elastase, 211
Nitrocellulose (filter paper), 130, 132
    for blotting techniques, 133–35
Nitrogen fixation, 225–27, 231
Non-coding sequences, 82
Non-coding strand, 30, 31, 33, 34
Non-selectable genes, 179
Nonsense mutation, 49, 50, 52
Northern blotting, 135, 138
Nucleic acids, structure of, 28–29, 70. *See also* DNA; RNA
Nucleotides
    DNA, 13–16
        base pairing of, 16–17
        cleaving of, 63–65
        in recombinant, 67–70
    dideoxy-, 88–90
    oligo-, 133
    of promoters, 145–46
    purine, 181–82
    RNA, 32–36
    structure of, 21–22, 28–29
    terminal, 63

O

Oligo (dT) cellulose column, 99, 114
Oligonucleotide probes, 133
Opérons, 157, 158
    lac, 143–45
    trp and repressible, 145
Operator, 143, 157

P

Palmiter, Richard, 177, 202
Particle bombardment, 223, 230
Pathogens, 211–12
pBR322 plasmid, 106–8
Pedigree analysis, 194–95
Penicillin, 59
    ampicillin, 106–7

Pentose sugar, 13
Peptide bond, 39
Pesticides, 224-25
Phages. *See* Bacteriophages
Phaseolin, 228
Phenotype
   definition of, 4-6
   mutations altering, 50-51, 52
   relationship of genes to, 26-27
Phenylalanine, 192
Phenylketonuria (PKU), 192-93
Philadelphia chromosome, 47
Phosphate group ($-PO_4$) of DNA
   removal of, in cloning, 127-28
   structure of, 13-16
Pilus, 120
Plants
   amino acids in, 227-28
   cloning of cells of, 221-23
   nitrogen fixation in, 225-27
   resistance to herbicides and pesticides, 224-25
   resistance to frost, 227
Plaques, 118
Plasmids
   characteristics of, 58-59, 114
   as cloning vectors, 105-6
      cosmid, 111-13
      pBR322, 106-8
      shuttle, 166-67, 169
      Ti, 222
      for yeast cells, 164-66, 168-69
   isolation of, 61, 132
   2 micron circle, 162-63, 168
   pSC101, 124-26
Plating cells, 121
Pneumocystis carinii pneumonia, 216
Point mutations
   in sickle-cell anemia, 23-25, 50-51
   types of, 49-50
Polyacrylamide gel electrophoresis, 85-87
Polyadenylation, 33
Poly(A) synthetase, 154
Poly(A) tail, 33
   in isolating mRNA, 98-99, 114
Polyethylene glycol, 163, 173, 186

Polymerization, 19
Polymorphisms. *See* Genetic polymorphism
Polynucleotide strands, 14, 15-16
Polypeptides, 23
   of hemoglobin, 24-25, 27
   synthesis of, 28, 29, 36-39
Polytene chromosomal puffs, 152-53
Positive regulation, 146-47
Post-transcriptional regulation, 156
Post-translational regulation, 156
Precipitated DNA, 60-61
Predictive medicine, 217
Pribnow box, 30, 146, 155, 158, 184
Primer DNA, 87, 88, 89, 100
Probes
   in diagnosis of human disease, 217-18, 219
   in screening genes, 130-32, 138
   types of, 132-37
Proinsulin, 156
Prokaryotic cells, 30
   expression of eukaryotic DNA in, 156-57
   gene expression in
      positive and negative, 146-48, 150-51
      lac operon control of, 142-46
      translational, 148-49
   isolation of DNA from, 61
   structure of, 56, 58, 71
Promoters, 143, 147, 158
   effectiveness of, 145-46
   interaction of CAP proteins with, 146-48
   in mammalian hosts, 184-85, 191
   RNA polymerase II and, 154-55
Pronuclei, 176, 177, 190
Protease, 60
Proteins
   activator, 146-47, 150-51, 154, 158
   difference between polypeptides and, 27
   fusion, 157
   histone, 56, 59, 70, 152
   primary structure of, 37
   r-, 149
   repressor, 143-44, 154, 157
   synthesis of, 36-39
Protoplasts, 163-64, 168, 221
pSC101 plasmid, 124
Pseudomonas fluorescens, 225

Pseudomonas syringae, 227
Puffs, chromosomal, 152-53, 154
Purine nucleotides, 181-82
Purines, 13, 16-17
Pyrimidines, 13, 16-17

R

Rabies, 213-24, 228
Reading frame, 37-39
    mutations in, 49-50, 52
Recognition sequences, 64-65, 66-67, 71
    Alu, 151, 179-80, 190
    Pribnow box and -35 sequence, 146, 158, 184
    Shine-Dalgarno, 39, 148-49, 158
Recombinant DNA molecule, 69, 71. *See also* Gene cloning
    blunt and staggered bases in, 67-69
    DNA ligase in, 69-70
Recombinant DNA technology, xi. *See also* Biotechnology
    hormones of, 205-7
    lymphokines of, 207-9
    therapeutic enzymes of, 209-11
    vaccines of, 211-14, 219, 228
Recombinant vectors, 126-28
Reference DNA, 78, 80
Regulator gene, 143, 144, 157
Relaxed plasmid, 106, 107, 114
Repetitive DNA, 151
Replica plating, 130
Replication of DNA
    chromosomal mutations in, 44-49
    cloning vectors in, 102-13
    gene mutations in, 49-52
    process of, 18-21, 22, 42-43
    rolling circle, 110, 113
Repressible operons, 145, 158
Repressor protein, 143-44, 154, 157
Restriction Fragment Length Polymorphisms (RFLPs), 83-85, 91
Restriction maps, 81-82, 90, 199-200
Restriction nucleases, xiii
    in cloning genes, 97, 102, 103-4, 105, 112, 113, 165-66
    in detecting genetic polymorphisms, 83
    in DNA cleavage, 62-66, 71, 74-75

    in DNA mapping, 81-82
    of pSC101, 124-26
    types of endo-, 66-70
Reticulocyte mRNA, 132
Retroviruses, 11, 100, 214
Reverse transcriptase, 100, 101, 214
Rhizobium, 226
Ribonuclease, 60, 61
Ribonucleic acid. *See* RNA
Ribonucleotides
    of genetic code, 32-36
    of mRNA, 32-33
    structure of, 28-29
Ribose, 29
Ribosomal RNA (rRNA), 29, 33, 40, 148-49, 155
Ribosome, 148
Ribosome binding site, 39, 148-49, 158
RNA, xiv
    in cloning genes, 97-98
    function of, 32-36
    probes to examine, 135
    production of, 30-32
    in retroviruses, 11
    structure of, 28-29
    translation of, 36-39, 40
RNA polymerase, 30, 31
    in eukaryotic gene expression, 154-55, 184
    in prokaryotic gene expression, 143, 144, 145, 146-47
RNase, 60, 61
r-proteins, 149
Rolling circle DNA replication, 109, 110, 113
Rosenberg, Steven, 208, 209
Rudman, Daniel, 206

S

Saccharomyces cerevisae, 162, 163
Same sense mutation, 49, 50, 52
Sanger dideox-DNA sequencing, 87-90, 91
Screening of gene libraries, 94, 129-32
    probes for, 132-37
Selectable markers, 104-5, 106, 124
    for mammalian hosts, 178-80, 181-83
Selective media, 124, 125, 126, 138
    HAT, 179, 190
    tissue culture, 173

Sendai virus, 173
Separation in transcription, 30, 31, 32, 40
Severe combined immune deficiency (SCID), 204-5
Sex chromosomes, 8-9
Shine-Dalgarno sequence, 39, 148-49, 158
Shotgun cloning, 118
Shuttle vectors, 166-67, 169
Sickle-cell anemia, 23-25, 50-51, 93, 200
Simian virus 40 (SV40), 64, 65, 81, 176
    as vector for mammalian hosts, 180-83, 184, 191
Small nuclear RNA (smRNA), 29, 40
Smallpox, 211, 212
Smith, Hamilton, 63
Somatic cell hybridization, 172-74, 190
    clone panel and, 196-97, 218
Somatic cells, 8, 51, 52
    gene replacement therapy in, 203-5
S1 nuclease, 100, 101
Southern blotting, 133-35, 138
Spheroplasts, 163-64, 168
Stable gene expression, 180
Stahl, Franklin, 18, 20
Start codon, 36, 39
Stem cells, 202
Sticky ends of DNA, 67-70, 71
Streptococcus, 76
Stringent plasmid, 106, 114
Structural genes, 143, 157
Substitution mutations, 49, 52
Substrates, 195
Subunit vaccines, 212-14, 219
Superovulation (in animal production), 228
Synapsis, 45

T

Tapeworms, 228
Target gene, 93. *See also* Gene expression
    selection and preparation of, 96-99
TATA box, 30, 155, 159, 184
Tatum, Edward, xiii, 26, 27
Tay-Sachs disease, 195
T DNA, 222
Templates, 18, 30

Terminal nucleotide, 63
Termination (in translation), 37, 38, 39, 40
Termini, 67
Tetracycline, 105, 124, 125
    resistant genes to, 106-7
tet^R cloning vector, 124
tet^s phenotype, 124, 125
T-4 helper cells, 215-16
-35 sequence, 146, 158, 184
Thymine, 13, 14, 16-17, 29
Thymine kinase (TK), 178-79, 190, 196-97
Ti plasmid, 222-23, 225, 230
Tissue culture medium, 173
Tissue plasminogen activator (tPA), 209-10
Transcription, 28, 29, 101
    basic steps of, 30-32
    in eukaryotic gene expression, 150-51, 154-56
    in prokaryotic gene expression, 143-44, 145-47
Transcriptional start site, 146, 155
Transduction, 123
Transfection, 123
Transfer DNA (tDNA), 29, 33, 34, 40, 155
Transformed cells, 96
    detecting, 123-26, 138
    selecting for mammalian, 178-80
Transgenic animal, 176-78, 190
Transient gene expression, 180
Translation, 28, 29, 36-39
    in eukaryotic gene expression, 156
    in prokaryotic gene expression, 143-44, 145
Translational gene regulation, 148-49
Translational start site, 149, 155
Translocation, 39, 52
    simple and reciprocal, 45-47, 48
Trisomy, 21, 44, 25
trp operon, 145, 158, 184
Trypanosomiasis, 228, 231
Tryptophan, 145
Tumor inducing (Ti) plasmid, 222-23, 225, 230
Tumor infiltrating lymphocytes (TILs), 208
Tumor necrosis factor (TNF), 208-9
Turner's syndrome, 44
Tyrosine, 192

U

Unexpressed genes, 139. *See also* Gene expression
Unique restriction sites, 103–4, 105, 106
Uracil, 14, 16–17, 29, 35

V

Vaccina virus, 213–14, 219
Vaccines, traditional and recombinant, 211–14, 219
Van Der Eb, A. J., 174
Vectors, 94
    characteristics of, 102–5, 114
    cosmid, 111–13
    expression, 118–19, 168
    identifying recombinant, 126–28
    phage, 108–11
    plasmid, 105–8, 164–65, 168–69
        shuttle, 166–67
        Ti, 222
    viral, 180–83
Viruses
    baculo-, 225
    human immunodeficiency, 11, 214–17
    mouse mammary tumor, 184–85
    phage T2 and life cycle of, 10–12
    Sendai, 173
    simian virus 40, 64, 65, 81, 180–83
    vaccines made from, 211–14, 219

W

Watson, James, xiii, xv, 15, 201
Western blotting, 136–37, 138
Wexler, Nancy, 84, 200
Whole virus vaccine, 212, 214
Wild-type mammalian cells, 180, 181, 183

X

Xanthine guanine phosphoribosyl transferase (XGPRT), 182–83
X chromosomes, 8–9
Xeroderma pigmentosum, 195–96
X-gal, 128

Y

Y chromosomes, 8–9
Yeast cells, 119
    complement mutations and, 166–67
    as eukaryotic host, 162–64
    expression of cloned genes in, 167–68
    homologous sequences and, 165–66
    plasmid vectors in transformation of, 164–65, 168–69

Z

Zygote, 8, 150, 176